W0229971

Training and development of technical staff in the textile industry

Training and development of technical staff in the textile industry

B. Purushothama

WOODHEAD PUBLISHING INDIA PVT LTD

New Delhi ● Cambridge ● Oxford ● Philadelphia

Published by Woodhead Publishing India Pvt. Ltd.
Woodhead Publishing India Pvt. Ltd., G-2, Vardaan House, 7/28, Ansari Road
Daryaganj, New Delhi – 110002, India
www.woodheadpublishingindia.com

Woodhead Publishing Limited, 80 High Street, Sawston, Cambridge,
CB22 3HJ UK

Woodhead Publishing USA 1518 Walnut Street, Suite1100, Philadelphia

www.woodheadpublishing.com

First published 2012, Woodhead Publishing India Pvt. Ltd.
© Woodhead Publishing India Pvt. Ltd., 2012

This book contains information obtained from authentic and highly regarded
sources. Reprinted material is quoted with permission. Reasonable efforts have
been made to publish reliable data and information, but the authors and the
publishers cannot assume responsibility for the validity of all materials. Neither
the authors nor the publishers, nor anyone else associated with this publication,
shall be liable for any loss, damage or liability directly or indirectly caused or
alleged to be caused by this book.
Neither this book nor any part may be reproduced or transmitted in any
form or by any means, electronic or mechanical, including photocopying,
microfilming and recording, or by any information storage or retrieval system,
without permission in writing from Woodhead Publishing India Pvt. Ltd.
The consent of Woodhead Publishing India Pvt. Ltd. does not extend to
copying for general distribution, for promotion, for creating new works, or for
resale. Specific permission must be obtained in writing from Woodhead
Publishing India Pvt. Ltd. for such copying.

Trademark notice: Product or corporate names may be trademarks or registered
trademarks, and are used only for identification and explanation, without intent
to infringe.

Woodhead Publishing India Pvt. Ltd. ISBN: 978-9-38030-821-0
Woodhead Publishing Ltd. ISBN: 978-0-85709-581-7

Typeset by Sunshine Graphics, New Delhi
Printed and bound in India by Replika Press Pvt. Ltd.

Contents

Woodhead Publishing India Series in Textiles

- **Fundamentals and Advances in Knitting Technology**
  Sadhan Chandra Ray

- **Industrial Engineering in Apparel Production**
  V. Ramesh Babu

- **Management of Technology Systems in Garment Industry**
  Gordana Colovic

- **A Practical Guide to Quality Management in Spinning**
  B. Purushothama

- **Modern Approach to Maintenance in Spinning**
  Neeraj Niijjaawan and Rashmi Niijjaawan

- **Performance of Home Textiles**
  Subrata Das

- **Fundamentals and Practices in Colouration of Textiles**
  J. N. Chakraborty

- **Science in Clothing Comfort**
  Apurba Das and R. Alagirusamy

- **Effective Implementation of Quality Management Systems**
  B. Purushothama

- **Handbook of Worsted Wool and Blended Suiting Process**
  R. S. Tomar

- **Quality Characterisation of Apparel**
  Subrata Das

- **Humidification and Ventilation Management in Textile Industry**
  B. Purushothama

- **Fundamentals of Designing for Textiles and Other End Uses**
  J. W. Parchure

- **High Speed Spinning of Polyester and Its Blends with Viscose**
  S. Y. Nanal

Preface

Technical staff in textile and apparel industry is the backbone for the industry to run successfully. The managements employ adequately qualified personnel from the point of view of technical knowledge and allot them the supervisory jobs, where as basic controls of the raw materials, men, machinery procurement, devising policies, etc., are kept by the top management.

There are number of colleges and educational institutions world over to impart technical knowledge; however, the actual work involved in the industry is more of managing the resources such as human resources, infrastructure and addressing the critical issues in supply chain, changing customer needs and expectations, production balancing, grievances or misunderstanding in implementing the systems and procedures, interpersonal conflicts and so on. Taking the best out of the existing infrastructure, machineries, men and the materials is the challenge faced by all technical staff. However, none of the technical universities, institutions or colleges are addressing these in there syllabus, and also it is not practicable for them to address these issues as they do not have adequate knowledge or experience.

Often we hear the top managements blaming their technical staff for the failures and losses the company is making, but what the top managements are doing to educate and train their technical staff in becoming efficient and effective supervisors is a million-dollar question. They expect the technicians to be self-made and produce results irrespective of all odds.

In industry there are different designations given to the supervisory staff, and unfortunately the word supervisor has lost its meaning. In the industry, the word supervisor means the lowest level of job done by a technical person, and numbers of companies are boasting themselves of eliminating the supervisors. One should understand that one can abolish the post of a supervisor, but not the supervisory functions. All the technical staff including the floor in-charges, functional heads, production managers, general managers and some times even the executive directors can be called as supervisors as they do the supervisory functions. There success depends on how they supervise their activities.

There are number of books and articles available dealing with technology and management separately, but explaining how these techniques could be used in the daily life of a supervisor is very few. This book is an effort to explain various aspects of managements related to daily working on shop floor by supervisory staff. I hope this book shall be a practical guide for the industry to develop their supervisory staff.

B. Purushothama

1
Technical staff development

1.1   The need for trained technical staff

The textile and apparel manufacturing is one of the oldest trades in the world and has innumerable units world over engaged in this activity. There is a high competition between segments and units and is bound to be high all the time. The success of a company depends on how competitively one can produce goods to the satisfaction of the customers at competitive price and deliver in time. Therefore, continuous developments can be seen in the technological aspects as well as in managerial aspects in the industry.

The industry which was highly labour intensive is gradually becoming capital intensive especially in the sectors of spinning, weaving, doubling and twisting, texturizing, wet processing, non-woven and technical textiles, where as it is still labour oriented in garment sector and decentralized sectors. Further the process of manufacturing has number of control steps to ensure that products are suitable for the ever changing requirements of the customers. Although, automation reduced the human activities, still there is a need for technical persons to monitor the processes to ensure that they are carried out as planned and the work practices are adhered to as needed to get the required quality and productivity. Such person is termed as a supervisor or a middle management staff. Anyone doing these activities is a supervisor.

The supervision is a very important activity in the industry to monitor the activities on person to person and minute to minute basis. The job involves both technical aspects and the human related approaches. There are number of technical colleges and polytechnics providing basic information on the technology, where as in the industry, the first job given to a technician is supervising the activities and monitoring them. The candidates completing the courses get their diplomas, bachelor or master degrees or even doctorates. The industry recruits them by seeing their marks and the way in which they answer in the interviews, but grumble them as not up to the mark when they start working in the industry. The colleges give lot of information to the students that can cover all types of possible jobs a candidate can take like that of working in a shop floor of an industry or working as a faculty in an educational institute or as a research scholar and so on. The college education gives the basic knowledge of technology but not on the practical activities a supervisor has to do in an industry. The colleges can give some hint on the roles and responsibilities of a supervisor, but cannot effectively train a supervisor. The reasons are many. Normally, one uses only 5% to 10% of what was learnt in the college while working in an industry, as the industry cannot have all the processes or machines taught in the college. Industry restricts itself to one or two specific processes and develops it as core competency. It expects their supervisors to concentrate on the process and the systems adopted and not to bother about other systems or technologies as it cannot give the results or they could not be adopted due to various reasons.

To get trained as a supervisor the candidate needs some exposure to industry working and the one giving training should have thorough practical experience in the industry. The trainer should not only impart skills to the candidates but also should improve their competency for solving their day-to-day problems by logical approaches. The services of such people in training the supervisors shall be very helpful to the industry.

1.2   Quality people – key to excellence

The globalization and opening up of world markets along with rapid technological developments are the salient features of the present economy. Severe competition between and within industries can be seen worldwide. Everyone is trying to be competitive. Installing the latest possible technology and introducing new products with special features are the normal strategies adopted. Huge amounts are spent for installing the laboratories, plant and machinery with state-of-art technology and highly qualified managers and sub-managers are employed to run the plant. The image of "quality people" and "quality company" are felt as very essential to attract the customers to be in the market and run the business successfully.

All organizations wish to be successful in their ventures and do a lot to get the reputation of a 'quality company'. The efforts include improving the conditions by adopting latest technologies, cutting costs wherever possible, recruiting people with high qualification and experience, offering good remunerations and packages to attract and retain competent staff, training and educating people, benchmarking and adopting systems that brought success to other leading companies, trying to develop a culture compatible to the requirements and impressing the customers and public by various means. People feel proud while working with latest machines and equipments. Customers feel that a company with latest technology and best infrastructure is capable of giving consistent good quality. The customers expect a consistent product at low price delivered at shortest notice because of the latest State-of-Art technology adopted. The management also ensure best possible raw materials and accessories (even by paying premium prices) to achieve the quality products. Inter company comparisons are made for quality and efforts to always remain on the top. But we see that in a number of cases the companies are struggling to survive. The members of the management are grumbling that even after investing huge amounts on new machines, employing highly qualified professionals by paying very high pay package, spending huge amounts for getting various certificates like ISO 9000, ISO 14000, SA 8000, etc., there are no returns. The market did not pay any premium for the quality goods made with latest machines, it goes by the lowest quotation and whatever earned is just sucked by the banks as interest and as taxes by government.

Only by having the state-of-art technology cannot help a company in succeeding in the competition, where as people with competency to manage the technology is more important. Incompetent persons can spoil the company even with latest technology, where as competent people can manage the company even with old technology, as they know the way of working with the available technology and taking best out of it. It is normally seen that the industry users depend much on the manufacturers for the maintenance of their machines including the scheduling, replacement of parts, overhauling and settings, whereas it should be the work of shop floor technicians, as they are the users. If shop floor technicians take interest and practice the tuning of the machines, they should be able to get the required quality and productivity. The fear of failure or of spoiling the costly machines make the management think of taking the help of machinery manufacturers, but by that, they are making their own people inefficient and useless. One should remember that unless he jumps into water he cannot learn swimming. One should be self sufficient to protect self and come out successfully.

In textile and apparel industry, technology is developing fast and the dependency on the human beings is being reduced by automation. The machine speeds are reaching high levels. Although, the attempts are made to reduce human errors, the men on spot needs to be more alert with new high speed machines than handling conventional slow speed machines, as the loss shall be high with high speed machines even for small mistakes or stoppages. Unless the persons are well trained and work with presence of mind, the investment for new technology shall be fatal. The role of technical persons become vital as they need to monitor the process and get the results as required. Therefore, it is needed to train the technicians on continuous basis depending on the changes that are taking place in the technology and the changing requirements of the customers.

1.3   Duration of training

Industry needs good supervisors, but is not in a position to spare them for long term training courses. The supervisors working in industry shall have observed number of things and would have a fair idea about their roles and responsibilities, and need not be given detailed lectures as given to college students who have not seen the industry. Therefore, short term courses need to be developed to train the supervisors, while understanding the working procedures of the mills in which they are employed or likely to be employed. The duration depends mainly on the existing knowledge and skills of the supervisor and the expectation of the management. There is a need for modular approach, rather than a general training for the specific needs of the industry. The duration of training might vary from 200 hours to 600 hours, i.e. 1 month to 3 months. The industry may collaborate with any of the technical training institutes dealing in textile and garment technology and get their staff trained in a systematic way and have tests to ensure that the candidates make efforts to learn, and implement what was learnt. Alternatively, where facility is available, one of the senior staff can train the supervisors as per the curricula suggested, but it is suggested to have tests conducted by an external expert.

1.4   Recognition

People shall be proud when someone recognises their work. Candidates undergoing training also shall be proud when they get a certificate or a memento for attending the training. A competent authority need to conduct a test at the end of the programme by a panel of experts, who were not involved in giving training. The successful candidates should be issued with a certificate that is recognised by the industry. The recognition shall make the candidates proud of their jobs and improves their moral. The industry can be decided on the competent authority depending on their critical requirements. It might be an educational institute, a research organization, and an association of professionals or their own staff. However, the certificates have more weight when given by an approved authority from the government, university or a professional body.

1.5   Training modules

Depending on the requirements of individual mills or the garment manufacturing units, different modules can be developed. The modules should be prepared considering the specific works to be done and the existing level of competency of the people on the job. The course curricula should contain the practical competencies, the underpinning knowledge to attain that practical competency, the tools and equipments needed the control points and the check points in each operation, the effect of various aspects on the quality and production, the critical quality requirements and the soft skills for the supervisors to manage their sections. This book gives general guidelines, whereas, it is suggested that the individual mill/garment factory prepare their own curricula depending on the actual machinery they have, the systems being followed and the prevailing culture.

In this book, discussions are made on general requirements of a supervisor, and specific modules are discussed relating to the technical staff of textile and garment industry.

Technical staff – the middle management

2.1   Introduction

Technicians play crucial role in managing the process of any organization. They are involved in the design and establishment of the process, coordinating with the people on the shop floor and making them understand, implementing the process, monitoring and correcting them. Therefore, the credit for success or failure of a process goes to the technicians on the floor.

The technicians are always under the pressure of producing the required quantity in time as per the agreed quality irrespective of the odd situations faced due to various factors relating to labour management, power shortage, maintenance lapses, non availability of critical parts, changes in climatic conditions, and sudden changes in customer requirements and so on. They have no right to give an excuse; instead they have to give results under all circumstances by foreseeing the problems and taking precautionary measures. They have been respected for this ability through out the world. However, unless one applies his mind and work out solutions for various problems faced, it shall not be possible for him to be successful as a good technician.

2.2   Roles and responsibilities of supervisors

The supervisor is the manager of his section and has the rights of achieving the sectional and departmental objectives by planning, coordinating, implementing and controlling the activities. The supervisor, who is a mediator between the top management and bottom line workers, have the following responsibilities:

- Understanding precisely the requirements of the customer and communicating to the workers producing the quality in the language understandable by them.
- Understanding the company capabilities, like process capability, human resource capability, funds availability, material availability,

infra structure availability, etc., and working out the plans to accomplish the production requirements.

- Understanding the ethical and legal requirements of the process for the products and services planned and monitoring to meet the requirements.
- Designing the product that is acceptable to customer and the society.
- Designing the process to get optimum output at minimum time and cost.
- Deciding the measures for measuring and monitoring processes and monitoring accordingly.
- Working out the detailed quality plans and the actions needed at different levels and ensuring the same is implemented as planned.
- Working out the production programme, implementing and monitoring them.
- Planning for the raw materials, spares and consumables and indenting for the same as needed.
- Procuring the required materials in time and ensuring their quality by proper inspection and testing.
- Planning the maintenance activities and implementing them to have minimum loss of time for maintenance while ensuring lowest possible failures.
- Tuning the machines as per the process designed while understanding the technical capabilities.
- Educating and training the men on shop floor on the production and quality aspects of the product and the process.
- Allocating suitable competent workmen for the skilled jobs and monitoring their works relating to quality and productivity.
- Maintaining harmony in the work place and creating a good work environment.
- Monitoring the process periodically to ensure its suitability and to have lowest wastes possible.
- Documenting the procedures adopted, actions taken on problems and deviations in the results, monitoring and ensuring following of procedures all the time.
- Reporting the activities suitably to superiors as well as to the next person taking charge, highlighting the deviations and special actions.
- Analysing the reasons for deviations in process and product performance and discussing with the superiors for suitable actions and taking actions as decided.
- Arranging for monthly, quarterly, half yearly and annual stock taking as per the requirement of the company and the section.
- Working out the realization of fibres, yarns and fabrics on periodic

basis referring to actual consumptions and stocks and the material actually realised.

- Maintaining harmony at work by adhering to company's policies, legal and regulatory requirements, while understanding the needs of stake holders and taking all the workers together to achieve the objectives and goals.

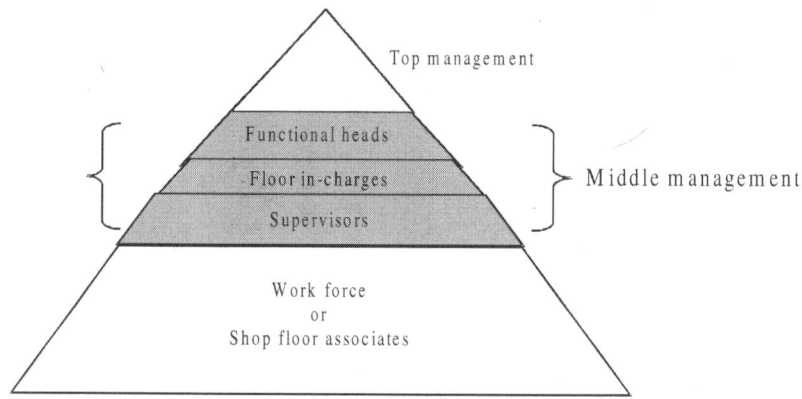

2.1 Hierarchy of management.

2.3   The organization structure

Any organization shall have three levels, the investors or the owners in the top referred as top management, the managerial and technical staff termed as middle management and the operators or workers working on the shop floor. The technical staffs do the work of understanding the top management requirements and policies and deploying them down the line. The supervisor shall act as a mediator between the two important layers of the organization, viz. top management and workforce (Fig. 2.2).

The middle management gets broad guidelines and procedures from the top, which need to be converted into specific instructions before giving to workforce or the shop floor associates. They plan the activities depending on the annual plans and the policies. They train the people as needed and review the activities on shift-to-shift basis and keep a track of production and quality.

2.4   Routine and special activities

The works in an organization can be grouped as "routine" and "special" activities. The routine is supposed to be done regularly with out any deviation. Normally, these works are allotted to people who are good

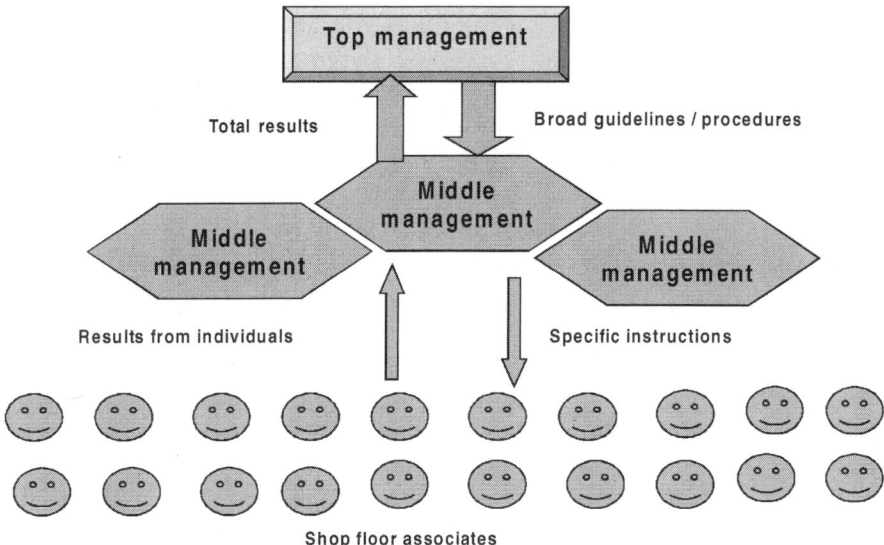

2.2 Communication by middle management.

followers and religiously follow the steps. The works include the recording of attendance, production, data of machine-wise production, quality, wastes, speed, efficiency losses, etc, labelling the products, house keeping activities, scouring of machines, replacement of lubricants, replacement of ring travellers, and so on. The Special activities require creativity and thinking and the jobs are non-repetitive in nature, for example some modifications on the existing machines, modification in existing systems, launching of new products, special trainings given for staff and workmen, fixing of new standards, etc. The technicians normally like special jobs so that they can show their capabilities, unfortunately, they forget that unless the routine works are done systematically, the special works done does not give results in large-scale productions, and in maintaining the systems. Therefore, the special activities should always ensure that the routines are not disturbed. The routine activities are the backbone of any successful organization, and the rigidity in following the systems gives the result, when the process is designed logically. A well-designed and followed routine works ensure the stability of the organization, and assures the quality and productivity all the time. Let us take some examples.

1. Reducing the imperfection in cotton yarn

The imperfections in the yarn can be reduced by various means, like introducing a better cotton with good maturity, uniform length and lesser trash, reducing the production speeds, increasing the saleable wastes at blow room, carding and combers, optimizing the settings, ensuring proper maintenance of all parts in all the machines, maintaining the required temperature and humidity, educating the workmen periodically for good house keeping and work practices, selection of correct travellers, spacers, etc, replacement of cots and aprons, keeping the drafting zone clean, selection of suitable draft combinations and hanks, etc. Among the above activities some are "routine" and should be followed with out fail. They are ensuring the proper maintenance of all parts in the machines, maintenance required temperature and humidity, ensuring proper house keeping, keeping the draft zone clean, replacement of travellers, cots, aprons, etc., in time and so on. The act of selection of suitable raw material, settings, speeds, waste levels are all "special" activities, and need not be done daily. Even in spite of giving good cottons, good machinery, proper settings, etc, if we do not keep the drafting zone clean or do not remove the wastes from the machines periodically as decided, we can never reduce imperfections. We must first ensure that routine works are done religiously.

2. Improving the productivity

The productivity can be improved by modifying the machines to take up higher speeds, reducing the waste levels, reducing the stoppages due to various reasons, making the hanks coarser in spinning machinery, increasing the speeds, improving the raw material, maintenance of correct humidity and temperature, etc. Among the above, the selection of raw material cannot be done daily. Similarly, increasing the speed is not a daily affair. But monitoring the stoppages, the wastes, the idle capacity are the routine works, which must be given priority all the times. If we do not manage these routine jobs, the modification of machines, giving a better raw material, etc., cannot give the required results. It might give a reverse result increasing the breakdowns, power costs, bad quality and so on, rather than increasing productivity.

3. New product development

New products are developed to have a competitive edge in the market. However, we should understand that once the customer accepts the product, we need to give that product in bulk, and it is no more remains as a new product. The following up of routine disciplines like maintenance, periodic checking of quality, following with workmen for house keeping and good work practices, maintaining the documents, following the codification system, monitoring of production, wastes, etc., are very essential for any product to remain in the market. There is no meaning of daily giving a new product to the market by making various permutations and combinations, unless we develop and implement a system of maintaining the product developed.

4. Fixation of work norms

The fixation of work norms is not a routine job, but need to be done routinely on a periodic basis. No work norm can remain permanent in any organization. The job of a technician is to go on observing the working and the developments that are being taken place elsewhere and workout the possible improvements so that neither the manpower nor the capital is underutilized. Sticking to the agreed work norms is a "routine" work, and it should be followed religiously. If we deviate in routine follow up of implementing the agreed work norms, we will never be able to implement new work norms, doesn't matter how logical and economical it looks.

The above examples explain the importance of maintaining the "routine" activities. The success of process lies in maintaining the systems rigidly and not in developing new systems. But we cannot close our eyes for new developments. Any new development, once accepted should be made as a routine, and not to be treated as special all the time.

To have an easy monitoring of the routine items, it is needed to identify the routine activities and list them. Once the activities are listed, the persons and the frequency of monitoring are to be identified. They can be grouped as monitoring, online monitoring, hourly, shift-wise, daily, weekly, fortnightly, and monthly. Depending on the operations, the works can be divided among people and make them accountable.

Monitoring the activities are not just overlooking and ensuring them as correct, but also proper documentation. One should be specific in designing a document; it should not be just filling up records by sitting at a corner, and writing every thing as okay. Recording the readings is more important and the comments can help in taking decisions, but can also lead to bias. These documents are meant for helping the industry to investigate and find root cause for the problems they face and not for satisfying an auditor or a manager. The technicians should avoid the comments as "Checked and found Okay", but insist on writing the actual readings. They need to decide on what is required to be documented, so that it shall be a useful document for further developmental activities. The technicians should always ensure that whatever work they do is adding value and not increasing cost.

2.5   Challenges to middle management

The middle management (Fig 2.3) staffs faces lot of challenges in their routine works. They are in continuous demand from the top for increasing production, improvement in quality, reduction in wastes and reduction in costs. They need to face daily problems like shortage of manpower, non receipt of material in time, frequent changes in styles, pressure of dispatch, employee grievances and so on. They need to act as a buffer between top management and workforce.

They need to take proactive actions to prevent problems. It is therefore, necessary to clearly spell out the tasks at different levels, their authorities and responsibilities and document them so that all are clear in their tasks. After understanding the job descriptions, the authorities and responsibilities, one needs to discuss on the methodology to be adopted in doing various tasks.

2.3 Middle management as a buffer.

2.6   Understanding the requirement of a customer

Understanding precisely the customer requirement is the first job in any process. Unless, we are clear in this that we cannot give the required quality of product and services. The customer gives his requirement and expectations in a form, normally as a specification, but shall express his real concerns only when there is a problem. The purchase orders do not contain the concerns of a customer. The technicians, therefore, should take interest in going through various feedbacks given by the customers. It is essential to segment the customers by the end-use of the product, the region and the diversification they have. Although, a particular customer has not expressed his concern for a particular product, we need to go by the concerns expressed by the customers in that segment and work for overcoming it. It is better that a technician visits the customer to understand the technical requirements, rather than depending on the descriptions given by commercial persons.

It is a fact that customers normally give minimum requirements in writing, but shall be expecting a lot more from the suppliers. For example, no customer writes that packing should be attractive, but shall reject the material if packing is not tidy. It is the duty of technicians to find out the impact of each parameter at the customer's end. They should discuss with their customers / seniors and prepare a list for requirements and educate the concerned shop floor people to monitor them. This exercise should be done customer segment-wise. Let us look into some examples.

- A weaver manufacturing fabric for defence end use is more particular about maintaining the average count at a certain level as the fabric weight have a tolerance, whereas a weaver making the same variety of fabric for selling in open market insists on slightly finer count so that he can get more length of fabric, as the fabrics are sold on length basis.
- In some countries the weather is always cold, where as in countries like India, it is normally hot. When yarn goes from a hot country to a cold country, its property changes, as the natural wax adhering to the cotton becomes brittle. Even within India, some places are always dry and some are always humid. The customer shall be specifying his requirements considering his weather conditions, whereas the supplier ensure those specifications in his working condition. Normally, the problem of weight shortage, excessive linting, loss of strength, etc., can be attributed to this. If we know clearly the conditions, we can always take that into consideration and design our products. The worst part is that the customers assume the supplier as knowing his conditions, and the supplier always thinks that the working atmosphere shall be identical with that of supplier.

The clause 7.4 of ISO 9001:2008, suggests the purchaser to specify his method of verification of the products, whereas clause 7.2 insists the supplier to identify the intended and unintended use of the product, and the unspecified requirement so that the customer's exact requirements could be met.

Another important factor is the language. Although it is said that the communication is in a particular language, the exact meaning of a particular word used might be different depending on the culture and colloquial language in practice leading to confusion.

The requirements of a customer may be classified as quality, delivery, price, after sales service, and response time. Studies by experts have shown that 90 to 94% of customer grievances are due to areas other than product quality. It does not mean that quality of product is not important, but the customer assumes the quality as granted. It should not be thought that technician is not responsible for complaints other than quality. The delivery depends on how we monitor the production and complete the lots in time as agreed with the customers. The price can be lowered if we are monitoring the wastes and wasteful activities, and ensure that we produce the goods at the lowest possible cost. The after sales service depends on the expertise of the technician who visits the customer and resolves the issues. The quick response also depends on how fast we respond to the enquiries by the marketing. A good communication with customer relating to the orders, specifications, deliveries, complaints, etc., are essential, and the technicians should take active part in understanding the real requirements of the customers.

It is also a fact, especially in textiles, that the market complaints shall be more when market is bad or when the rates are increasing. However, the technicians should work to eliminate those problems also, so that they can hold the customers with them for long.

There are various methods of understanding the customer requirements. Getting the competitor's samples and discussing on the improvements required is a better method rather than on discussing on the specifications and tolerances. We should know exactly that what is the impact of different parameters.

Sometimes, we need to manufacture the products with out a firm order, and they shall be sold in open market. In such cases, we should target a segment and go by the feedback given by the majority of users.

The real task of a shop-floor technician is in translating the customer requirements into in-process specifications. This is discussed in length in designing a product.

2.7   Understanding the company capabilities

Understanding the company capabilities and communicating precisely to market can reduce the friction between the customers and the marketing personnel. Please remember that we are not politicians giving assurances to voters to get elected, we are a part of a supply chain, supplying products to our customers, who in turn shall convert them and supply further. If my customer fails because of my mistake or false assurance it is a loss to our company. It is therefore, very essential to understand our capabilities and limitations, and work on improving them. The company capabilities can be studied in the following manner:

- Process capability, i.e., whether the process adopted can produce the product as per the requirement of the customer in terms of quality, cost and volume.
- Human resource capability, i.e., whether our people are capable to produce that product as required by the customer and give after sales service as required.
- Finance, i.e., whether we can afford to hold stocks of raw material, packing materials, finished goods, spares, etc., so as to ensure un-

interrupted supply to the customer as per the requirement.

- Environment, i.e., whether the present social environment is supporting our business with the customer. This might include the trade regulations, pollution norms, political issues, power supply situation, draught or floods affecting the working, season affecting the working, etc.

The roles of technician is very important in working out the capabilities and informing marketing from time to time so that they do not give false assurances to customer and lead to problems later. There is no meaning in giving false assurances just to sell one consignment, as by which we might loose the customers permanently. The clause 7.2 of ISO 9001:2008 insists that before entering into a contract the supplier ensures that he has a capability to meet the customer's requirements. This clarity in business helps the customers to have confidence on suppliers, and they can come forward to work together.

In order to have quick information, it is essential to collect data of each machine and each person and go on entering the data to get trends. We should be able to know as to which machine and which worker are good for which product, and also a clear knowledge as to which machine or man is not suitable for this product. This is more important as the market complaints are mainly due to such machine or workers.

The technicians cannot keep quite after identifying certain process or machine or man as not suitable for a particular product or customer. He has to work for improving the situation, as the management wants to supply that product to that specific customer, as it is profitable. The technician has to work for improving the situation and make the men, machine and process capable of supplying that material to customer. He shall have to continuously work and give proposals to top management and improve the systems. Remember, others cannot do this job.

2.8 Understanding the legal requirements of the process

It is very important to identify the legal implications of a process and take suitable proactive measures to adhere to the requirements. This includes safety norms, pollution norms, ban imposed on certain chemicals and packing materials, ban on employment of child labour, non employment of ladies for certain jobs and also after certain hours of the day, timings of operating a siren during the day, various acts relating to factory, insurance, welfare, licensing of process, products and capacity, adhering to imports and export formalities, filing of statutory returns to the notified authorities, providing certificates of various nature, providing information on the packages relating to the producer, product, etc., getting the audits done of specified operations by certified authorities, issuing notices in time and so on. It is the duty of the technicians to ensure that the legal requirements are met, as in case of any violations they are the one answerable. One has to take the help of government recognized bodies for identifying the real requirements as per the law in force. In India, the textile commissioner's office provides information to the mills through circulars. A number of mill owners associations are doing a job of regularly collecting the information on the changes in regulations and send circulars to their member mills.

Among the legal requirements, some are related to safety of workmen and the community and others on the economic aspects. We cannot take a chance in ignoring safety aspects, as it can hit us also. The pollution might not hit the company directly, but as it hits the community around the company, shall indirectly hit the company. The economic aspects have an impact on the country's economy, by which the development works shall be hampered, and as an industry, we also shall be deprived of the benefits if the sufficient funds are not generated for the government for taking up the developmental works, whereas other countries shall be moving forward with developments.

One should always note that the regulations are normally made for the benefit of the industry and the community, which shall protect the men and investments. However, people normally feel these as hurdles in their progress and try to find a short- cut. Of course, there might be some regulation, which might not be in the interest of our company, and we need to resolve those things through proper channels and not by finding a short-cut. A short-cut shall make us a defaulter and shall be punishable under the law.

Make a list of applicable regulatory requirements for your section of operations, and indicate from where you can refer for the standards and norms. Have a record of fulfilling the requirements, which is normally called as a compliance record of your section so that you can monitor when some work is pending. Inform the concerned from time to time in case of any deviations found in order to take corrective measurements in time.

The safety aspects cover not only the process but also the product. One should ensure that products manufactured and supplied are safe for use at customers end. The quality of packing, the dimensions and weight of packages, handling systems, preservatives used, mode of transportation, other materials stored and transported along with these materials, insurance covered, etc., are all to be considered and monitored by the technicians. In some cases product safety sheets are to be signed by the head of production operations depending on the product. Normally, this applies to the wet processed materials, materials specifically made for medical end use and for use by babies, and when the material is supposed to be used in a high-risk zone where failures can be fatal.

It is essential for the technicians to follow up with the concerned HRD personnel relating to the periodic check up of their employees for medical fitness depending on the nature of work allotted to them. They should identify the risks in the job that can lead to sickness and hazards and take precautionary measures in time, find ways and means for reducing those risks by properly designing the process, and monitoring it.

2.9   Designing the product

Designing the product is considered as the sole responsibility of technical personnel, although, they depend on the inputs from other persons like marketing, purchase, human resource, costing, etc. The process of designing starts from the selection of designing stage. The technician knows by altering what, he can achieve which type of effect. His knowledge and experience can always help in identifying the stage at which designing could be made more effective. Therefore, the technicians are expected to have calm thinking and able to recollect from their experiences.

One of the important works of a technician is to provide correct inputs for designing, including the process capabilities, process suitability, effect of the process on other products and services, the human resource competency requirement, various data of earlier trials and developments, present limitations, availability of required raw materials and spares, technical know how, skills, infrastructure, etc. Product designing is one of the toughest jobs, as it has to meet all the requirements of customer, regulations, etc., and should be produced at the cost without hampering other activities. If the inputs provided are insufficient, then the designer cannot design properly. Therefore, the technician needs to have an access to relevant data, capture them and provide to the designer as needed.

While designing a product, the technician has to imagine his product as working at the customer's end, and think of the probable failures, which can affect the performance. Workout the loss a customer has to face in case of each failure and workout the priority in terms of risks and cost. One can afford to have breakages on a loom or a knitting machine, that can increase the cost of operation, but a snap in a rope of a hoist, or a snap in the conveyors can become fatal, and even a single failure is not acceptable. Similarly, there are different weightage for the problems a customer faces depending on the risk involved. Some mistakes can lead to closure of business itself whereas some are tolerable. It is therefore, suggested to make a list of priorities of problems and design solutions and monitor those areas where the risks and loss are high. This technique is referred as FMEA, i.e., Failure Mode Effect Analysis. To be successful in this aspect a technician needs to exert and understand the working culture at customer's end, and the type of mistakes that can be done at customers end and prepare a suitable fool-proofing device in the design that is being made.

The success of a designer depends mainly on how he makes the product fool-proof in spite of the mistakes done at customer's end. He should plan for all the three types of mistakes, viz. inadvertent mistakes, technical mistakes and conscious mistakes.

- The inadvertent mistakes are not regularly happening, and the persons are not interested in making a mistake. For example, writing a wrong identification on a product, skipping a line in the instructions provided, not observing a defect in the running batch or process, etc. These mistakes are not regular, but happen at times in spite of specific trainings and instructions given and employing good experienced workmen. The only method to avoid this type of problems is to reduce the dependency on the people in the process and make them fool-proof.
- The technical errors are those, which normally happen, especially with certain known persons, although they are not interested in making mistakes. This is mainly because of lack of confidence. This can be corrected by suitable training by which their skills can be developed and confidence is brought. Fool proofing of the systems always helps.
- People do the conscious mistakes, even after knowing that they are doing wrong. This includes the cooking up of figures, hiding facts, pushing materials of non-standard or different lots just to make up the packing, running a defective machine, stopping the safety gadgets to avoid stoppages, etc. Some of the mistakes may be worker driven and others might be management driven. Whatever may be the reason, the blame comes on the technician who designed the product or the systems. It is always better to design the system to be fool-proof. In order to make it fool-proof, one needs to have analysis of all the previous failures and their root causes. This can help us in devising suitable fool proofing.

2.10   Designing the process

The role of technician in process design is very crucial, as the success of a product in the market, the economy of the company, etc., depend on the quality of the process. A good process aims at highest productivity with the required quality at lowest cost. The process design involves understanding of the strength, weakness, opportunities and threats of each process, the interaction between processes, the linkages, balancing for production and quality, deciding of the machineries, speeds, settings, the recipe, the production target, the controls and checks at appropriate places, reviews to be made, targets to achieve, wastes generation and disposal, generation and utilization of by-products, trainings to be given, monitoring the training activities, deciding of the documentation, planning for storage of in-process material and the finished goods, identification system for materials, system of handling, stacking, preservation and dispatch, and so on. The product can be produced as required by various combinations of processes, but the one, which suits us best, should be decided considering the resources available. The technician has the knowledge of the activities involved in process designing and can forecast the impact of various parameters in the process design on the quality and productivity. Therefore, the technician needs to be always on his toes and improve his knowledge.

A process design requires a number of exercises of collecting the data of present system and analysing the situation, working out various process combinations by designing the experiments, formulation and verification of theories for the cause and effect of process changes, balancing of the resources and infrastructure available to get the best result, identification of training needs and planning for providing training synchronizing with the implementation of process change, etc. The normal input for process design is:

- Customer's requirements (quality, price, delivery schedule and service).
- The machinery available, their condition, limitations, and their availability.
- The raw materials and other accessories availability.
- The competency of men required and available.
- The infrastructure required and their availability.
- Present process and the limitations.
- Data on process combinations tried earlier and adopted for various purposes.
- The quality complaints and feedbacks normally received for products and services using similar processes.
- The regulatory requirements relating to noise, dust, pollution, light, safety, etc.
- The processes normally adopted by others for similar product and services.
- The interaction between processes and their limitations.
- The availability of technical know-how.

The output expected form process designs are the raw materials to be used, the machinery to be used, the speed, settings and other process parameters to be adopted, expected production rate, expected quality level, tests and inspections to be done, reviews to be made, records to be maintained, men to be engaged and trained, precautions to be taken, education to be given to users and customers, the costs to be incurred and so on.

The technician should keep his information bank always open, and ensure that it is active. Knowledge cannot be put in a fixed deposit. It attracts higher interest in current deposits and loses value in fixed deposits.

The main activity in process designing is educating the people on the spot for successful implementation of the designed process. Here the role of a technician is very important. In a number of cases it is seen that written documents are provided for people on spot to follow a system, but this is not effective. Only by reading we cannot understand the systems. It requires guidance, practical demonstrations, on the job training and experience exchange. How much we learn by different means can be summarized as follows.

- Only by reading – Up to 15.0%
- By taking guidance – Up to 30.0%
- By seeing – Up to 50.0%
- By doing – Up to 85.0%
- By sharing knowledge and guiding others – Up to 90.0%
- By practicing through out life – Up to 99.0%

Therefore, the technicians should work practically and demonstrate to the people the working as per the new process designed.

2.11   Deciding the measuring and monitoring of process

The measuring and monitoring of the process is very essential to achieve the results as anticipated. What is to be measured and how it should be measured is a pure technical job, and a real technician can only decide. We should measure those areas where we can do something. There is no meaning in measuring where we are not taking any action. Measuring for academic interest is not the objective of a technician working in a mill. We need to prepare the list of what are to be measured, and what we intent to do with the data. Once this is clear, we can workout the accuracy required in the data collection. Depending on the accuracy requirement, the measuring tool can be decided or designed.

The technicians should first prepare a relationship diagram indicating the cause and effects, which helps in designing the measuring devices. He can always plan for integrating the systems to source data for different applications.

Once the data is made available for the concerned persons, the act of monitoring shall start. The technician is important, as he knows the side effects of different monitoring actions. For example, to reduce the cost of manufacturing, one might think of reducing the picks per inch in a fabric, or reduce the size pick-up etc. A technician knows that what happens with this type of decisions. He knows the value of the work he is doing, and where necessary, he shall reduce the speeds, increase the wastes, but still brings profit by maintaining a good quality and consistency in production to help in winning the customers. A good technician knows the efforts required and the result that could be got by monitoring, and shall suggest the area, which can give maximum benefit. There is no meaning in controlling the stationary, lubricants, etc, but one need to control the realization of costly raw material, utilization of the plant capacity, etc. It means the knowledge of costing at each process element. Without proper knowledge of cost and effect, one cannot design the monitoring systems effectively.

The critical monitoring areas of a textile mills are normally the raw material costs, colours and chemicals, power consumption, machine utilization, operating cost per unit of production, rejections after manufacture, process stocks, production per unit, machinery down time for various reasons, material handling expenses, packing and forwarding expenses, market feedbacks, customer satisfaction, adherence to the set procedures and systems, morale of employees, safety and environmental protection activities, training of workmen for good work practices, etc. In all the above jobs, the technicians take an active role.

The act of monitoring involves studying the trend, identifying their root causes, forecasting the change, taking precautionary action, re-devising the measuring system, involving the people by educating properly to maintain the process and finally ensuring that the results are obtained as per planned efforts.

The monitoring process may be on-line or off-line depending on the process. Because of development in technology and information systems, almost all machines have one or the other online information providing systems. The basic problem is that the technicians do not find time to see the information. If we have no time to see the reports, then there is no meaning in spending money for such system. The real worth of a technician lies in how best he makes use of the information available. Unless we device systems to prevent bad material from moving forward, there is no meaning in our systems. The technicians should give priorities for controls. There is no meaning of controlling every thing, as it adds to the cost, and does not give any benefit to the customer. For example, keeping very close settings in autoconers is useless unless we know how to produce a good yarn. 100% inspection of finished material does not serve any purpose, unless the process is monitored effectively. Only technician can think and do this.

2.12   Working out the quality plans

The quality plans prepared consider the objectives of the product, the process capability and the quantity to be produced. Depending on the process capability, the machines, men and other infrastructure are allotted, and the testing and inspection plans decided. As a technician knows about his machines, he can foresee the problems, and decide on the monitoring to be done and the person to handle the machine and process. He can workout various combinations for balancing the process, by altering the speed, feed, delivery, etc, so that he can get optimum utilization of plant and machinery. He may plan for coarser hank in back process for a coarse yarn for a lower end use, whereas for a higher end use, he concentrates on the final product quality rather than just production of back process. He allots the machines considering the conditions and not by just numbers. For a non-technical man all ring frames are same, but for a technical man each spindle is different. Whenever, he wants to conduct trials or wants to make comparisons, he maintains the same spindles. Similarly, he knows the value and worthiness of each and every operation he does, and hence, he is the most suitable man for making quality plans.

A well-designed quality plan makes the job easy for the shop floor supervisors, who have to just follow the process as per plan. The junior supervisors should be alert in identifying the variation in process while implementing plans. The role of junior supervisors is to provide the required data for planning, like the problems in each machine, the quality levels achieved machine wise, the problems faced during the shift working relating to working, absenteeism, humidity controls, shortage or excess of process observed during working, etc. They should observe the process carefully and identify the potential problems so that suitable precautionary actions can be taken.

Analysis of the market complaints and their causes are to be explained to each and every workman involved, and their suggestions are to be taken to overcome the same. This should be considered in quality plans. The inspection and test plans prepared in the quality plans should have a direct link to the market requirements and feed back.

Once the plans are prepared, the concerned people in the organization like purchases, HRD, marketing, and quality control, etc., are to be communicated suitably so that they can workout their plans and act. Here again the technician plays a vital role.

2.13   Working out the production programme

The most important routine task of a technician is working out the production programme of the complete mills by balancing the production of each variety to meet the required delivery schedules. In quality planning, we discuss the activities relating to one particular contract, whereas here we are seeing the complete mill as one. We have limited machinery of different capabilities, and have to allocate them to have a balanced production. Here it is not only the balancing of production; it also includes balancing of men. As one contract is completed, the fresh contract to start shall be with a different variety, and hence the balancing of production is a daily affair, and hence is a routine job.

There shall be pressures from marketing to supply one particular variety early, because of which we may loose the productivity at other places, as the time required for different varieties are different. It is true for all the processes in a textile mill. In spinning area the production per spindle differs depending on the count and TPI, in warping and sizing by set lengths and creel size in addition to count, in weaving by the weave, the picks per inch and count, and in wet processing by the different shades, finishes, etc, in knitting by the course and wales per inch and the knitting pattern and in garment manufacture on the designs. Hence it is not possible to have a fixed pattern unless that mill produces fixed varieties all the time with out increasing or decreasing any of the products. It is not an impossible task as a number of mills are successfully managing to work with out changes; however the majority of the mills are flexible.

The balancing of production has to be done in all the shifts depending on the availability of men, machine, completing time of lots and contracts, stoppages due to unforeseen reasons like break downs, power failures etc, and disturbance in working due to various reasons, quality rejections leading for reworking, and so on.

2.14   Planning for the raw materials, spares, consumables, etc

The planning of the materials required for production is one of the major works of a technician. He need to plan in advance and indent for the required raw materials, consumables like ring travellers, wax, colours, chemicals, etc., maintenance accessories, lubricants, spares, packing materials, etc., which all depends on the customer requirements. However, planning is done much in advance before getting the contracts for manufacture. It means the technician should have data on the trends in customer requirements, which some times shall be seasonal. A timely planning and placing orders in time only help the mills to run smoothly.

2.15   Procuring required material in time

A planning has no meaning unless it is implemented. This holds good for procuring the required materials like raw materials, colours, chemicals, spare-parts, packing materials, etc. The technician knows the manufacturing programme and accordingly should follow up with the concerned purchasing people to get the material in time. The commercial people at purchasing cannot understand the importance of the items ordered, and they treat all as same, and for them the yardstick is the cost.

Once the material is received, the concerned technician should verify the materials and approve them for use. Unless the quality is acceptable, it should not be used.

The technician has a role in vendor analysis. Different weightage for quality, price, service and delivery are to be decided by the user technician and not by the commercial person. The weightage depend on the role of that material in our operations, and one should not refer to the weightage given by some one at some other mill.

The technicians can suggest alternate products in case of non-availability of certain ordered material. They cannot delegate this responsibility to commercial people.

Writing correct specifications of the materials to be ordered is the responsibility of technicians. The specification should be complete. In case of spares, we need to write the make and year of the machine, the part catalogue number, the drawing if any, and give samples if possible and feasible. In case of chemicals, we should demand for safety data sheet from the suppliers. Wherever applicable, the national or international standards are to be referred, for e.g., IS / BS/ EN/ ASTM/ ISO. He should also specify the method of inspection adopted for the received materials before they are approved for use.

The technicians should decide and prescribe the method of storing the received materials depending on their properties, frequency of using, safety aspects, etc.

2.16   Planning the maintenance activities

Maintenance activities are one of the crucial activities handled by technicians. The activities include understanding the machine features, identifying the critical parts, understanding the tuning operations of the machine, studying the life of various parts, forecasting the spares requirement in time, dismantling and reassembling of the machines for the purposes of scouring, cleaning, overhauling, etc., erection of new machines, shifting and re-erection of existing machines, ensuring breakdown free performance of the machines at optimum speed and productivity.

The frequency of maintenance activities are to be worked out by studying the condition of the machines and the working conditions, and not referring to any recommended figures of either manufacturers or research organizations. They can be used as a broad guideline, but we need to workout our programme depending on our machine conditions, working conditions, quality demands, customer feedbacks, break down history, the speeds, the product being produced, etc.

The maintenance job is a highly skilled job ant it need dedication, any lapse can result in bad quality and breakdowns, and finally reduces the working life of the machine. The work requires lot of concentration and determination. The maintenance works should never be postponed to achieve higher productions. While planning maintenance, precautions are to be taken to have minimum stoppages of the machines by planning the required parts, accessories, men, etc., in advance, and combining related activities.

The maintenance workmen should have undergone required training, and the knowledge of the machine. They should have the knowledge of basic tools and their application. A technician needs to train the workers and monitor the works. He himself should be able to demonstrate the work if he wants results.

It is always better to avoid opening of machines, as every dismantling and re-erections are potential problem of bad reassembling. It is always better to monitor certain parameters and study the trends. Once the trend shows the signs of deterioration, then only open the machines. The indicators selected should be such that they give an indication well in advance before the machine becomes bad.

Safety needs maximum importance during maintenance. The correct tools are to be used to avoid damages to the machines and men. Proper precautions are to be taken to protect the people working on the machines during maintenance. The electrical connections are to be completely removed otherwise; it is required for the maintenance operations itself. Proper signboards and locking system are to be made to prevent the machines from being started by mistake while maintenance operations are going on.

A deep knowledge of the lubricants is must for the maintenance man, as an improper lubrication can reduce the life of the machine, increase power consumption, and create problems of staining. The system of storing the lubricants and protecting them from being contaminated with dust and water vapour is very important.

A good maintenance man knows the role of each and every part provided in the machine, and does not allow any part to miss or to be bypassed unless an in-depth study is made. He allows modifications only after ensuring that there shall be no adverse side effect.

One of the very important tasks of maintenance is to maintain the maintenance equipments, and get the measuring and monitoring systems periodically calibrated. The tools and equipments needing calibration are the gauges of different types used for setting the machines, the temperature indicators, the pressure gauges, tachometers, volt-meters, multi-meters, measuring scales, weighing balances, eccentricity gauge, spirit levels, etc. The condition of the tools like spanners, wrenches, wises, hammers, screw drivers, cutting pliers etc., should always be in good condition or else they might slip and create more damage.

Wherever pressure vessels are used, care is to be taken in bolting all the leaks, verifying the strength of the body and getting the servicing verified by competent authorities.

The maintenance work should help in keeping the area clean and tidy. One should avoid spilling of oils, grease, wastes, water, chemicals, etc., on floor where maintenance activities are being carried out.

The technicians should monitor the quality and productivity by taking studies before and after maintenance and keep track of each machine by documenting the history of all changes done machine-wise to help in planning and taking preventive measures.

Please remember that a new machine can give good quality and production for some days, but shall be the cause for major problems if not maintained well, but a well maintained machine can never give problem, and the company shall be safe.

2.17   Tuning the machines as per the process design

Maintenance helps in preventing the breakdowns and bad quality due to worn out parts. Proper tuning of the machines, that is, setting of process parameters like speeds, setting, pressures, drafts, temperatures, timings of various motions, etc., are very important to get the correct quality and productivity. This is the prime responsibility of technicians. Even with latest machines we can produce worst quality if we do not know the parameters to be set. A deep study is required and one has to refer to previous settings, speeds and other parameters before making any change. If you feel change in parameters is essential, conduct study on a small scale, and change. Do not change all machines at one stroke. After every change, take sufficient studies to ensure that the changes have given the required results, and you are confident of continued results. The documentation of process parameters should be done only after successful establishment of the parameters.

Before tuning the machines, it is essential to ensure that the machine is in a good condition and do not have eccentric, vibrating or worn out parts, as they shall never give the correct results. It is the duty of the technician to ensure it.

2.18   Educating and training the men on shop floor

Educating the people on shop floor is essential to ensure that the process works as per plan. This includes giving information of the process, product, quality requirements, the customer feed back, the production expected, the precautions to be taken, the checks to be made, role of each man in achieving the target, and the general discipline of the company and the industry. The technician plays a major role in educating the workmen on the aspects of process, and he is the only man considered to be competent in doing that work. Non-technical people cannot explain and demonstrate the process to the workmen giving full information and clearing all doubts. Workers always respect and listen to one who can work with them and guide them when there is a problem. Practical demonstration of a system is always welcome, compared to classroom lectures or written instructions. Explaining in the language of workmen is required, and not showing our command on a foreign language.

2.19   Allocating the competent workmen for the skilled jobs

Just allocating a man for a job never helps unless he is competent persons for the job. It is the responsibility of the technicians to prescribe the minimum competency level for job. The competency is expressed in terms of education, skill, maturity and training. One who is having the required knowledge of the job, required skills, required maturity for taking decisions, and has undergone the required training to do the job is called as a "qualified person". We should not confuse with the university degrees or diplomas when we talk of qualified persons for a job. The university degrees or diplomas give a confidence in the employer that "such and such a person has passed such and such examination, and hence he is likely to have the knowledge of such and such a process". However, this might be a myth, as we have seen that a number of university-qualified people are not capable of doing the jobs compared to so-called uneducated people. Here, the most important thing is the personal capability and not a certificate. The technicians on the spot should observe the people working and judge their capability, and decide as to where they could fit best. A proper allocation of job reduces half the burden of the technician, and he can concentrate on other activities rather than following up with the people.

The process of identifying the minimum competency requirement involves breaking up of the job to small elements, and then writing the requirements. Here is an example of a ring frame tenter. The jobs done are creeling the bobbins, piecing the ends, cleaning the drafting zone, removing the pneumafil wastes in time, identifying the rouge spindles and informing the superiors, identifying any deviations and informing the supervisors, etc.

If we consider the work of creeling the bobbins, the tenter should have a knowledge of the count running on his machine, the colour codifications used in the speed frames, should have the practice of using a bobbin holder, should be tall enough to reach the top of the ring frame creel for keeping the bobbins or to remove the bobbins, should have the knowledge and practice of lifting the top arms, drawing the rove through drafting zone, etc. His education level should be sufficient to read the count board. His eyesight should be good to identify the different colours used for codifications, and to identify the spindles where bobbins are exhausted.

The act of piecing involves identifying the broken ends, stopping the spindle by his fingers, taking the end through the lappet hook and the traveller, feeling the tension of yarn before piecing, exactly putting the tail end of broken end to the front roller nip, etc. Here he is supposed to replace the traveller if missing or burnt, remove the lapping if any, collecting and keeping the bonda wastes at suitable location, etc. For this act the eyesight, finger dexterity, hand movements, and a good practice of piecing the ends is very important. He should be able to identify the traveller number and its suitability for the count he is working. His practice of sensing the tension of yarn can tell the history of the spindle. He can recognize worn out rings, worn out traveller, worn out lappet hook, worn out separator, eccentric spindles, less oil in bolster, loose tapes, improper pressure in top arm, etc. This skill cannot be taught in a college, but has to come by determined working and continuous application of mind while doing the job. A technician standing by the side of a ring frame and not having a practice of piecing can never be able to understand the problems in the machine as seen by a good tenter.

Cleaning of drafting zone is done when the ends are working, and the skill of a tenter lies in cleaning the area without cutting the ends and without introducing additional imperfections. Concentration in the work, skilful operation of taking a small stick and removing the embedded fluff from arbours, aprons holding brackets, necks of fluted rollers, pneumafil mouth pieces, side of traveller clearers, etc., requires lot of patience in doing the job. His knowledge on the role of each part in the quality and production of yarn plays an important role. Unless one knows the precision of the parts, he cannot develop that skill of handling them gently.

Removing pneumafil wastes in time might look as an unskilled job, but an experienced tenter shall be fast enough to sense the urgency and remove the wastes in time. The time he keeps the door open has an impact on the suction pressure. However, the new machines which have rotary system for continuous waste extraction does not depend on the skill of the operative.

Identification of a rogue spindle is not a separate job, but is a by-product of the skill and experience. As tenter feels tension and sound of each spindle during piecing and taking rounds, recognizes rouge, which is different from others. His maturity in deciding a particular spindle as rouge is very valued. An inexperienced technician insists the tenter to run the spindle as the yarn works on that, but a matured one shall see the quality of yarn produced and its effect on the total lot. A good technician recognizes the maturity of a tenter fast. Similarly, a matured tenter can identify the vibrations, worn out bearings, dry bearings, wrong meshing of wheels, deviations in temperature and humidity, change in cotton component, change in twist, speed, etc., and gives feedback to superiors in time.

By studying all the acts of a good sider, one can visualize the minimum competency required for allotting the job of tenter to a person. Similar is the case for any job. You just split the jobs into elements, and workout what is expected out at each step, and to achieve that, what type of man is required.

Only by writing the minimum competency level, the job of technician is not over. He should identify the existing competency levels and group them in different categories, and plan for suitable action like training, job transfer, job modification, providing assistance, etc. The action taken should help us in achieving our targets.

It is a normal practice to prepare skill matrix of tailors in a garment factory according to the various types of machines and stitches they have mastered. The tailors are put in the batch depending on the complexity in the style.

2.20   Monitoring the process periodically to ensure its performance

While monitoring the process to ensure its suitability the technician verifies its adherence to the planned system and takes suitable action in case of deviations. He shall have his own checklists or a checklist provided by his superiors, and verifies the activities. He shall not be able to check all, but shall have priorities to cover the vital activities. Depending on the changes done on that day or in the shift, he shall change his priorities. This includes production rates, wastes generated, quality of material in process, stoppages, house keeping, material flow, material handling, packing and storing, human productivity, team working, information gathering and maintaining records, maintaining the discipline while working relating to channelisation, colour codification, material storing, labour allocation, grievance handling, etc.

The monitoring of process includes providing the required work instructions, ensuring suitable working environment like temperature, humidity, air circulation, lighting, noise level, dust level, control of discharges, etc., following up with people to ensure that they are working as per plans, the production is coming as per the plans, the quality levels are maintained, the safety and regulatory requirements are fulfilled and taking suitable action in time for the deviations observed.

2.21   Documenting the procedures, actions and the results

The results got today should not be the end, but we should get the results continuously. A good technician shall document the happenings systematically, so that he can identify the reasons not only for failures, but also for the successes. The documents should clearly reflect the works done, actions taken, changes observed, reasons investigated, the root cause analysis made, etc. Just filling the pages cannot be called as a record. A record is meant for taking suitable action and hence, there should be some system of easy access to the required information by way of proper indexing. The technicians should decide on what is to be recorded, who should record, when it should be recorded, and where it should be kept. It is necessary to train the people on method of collecting data and information and recording them, so as to avoid mistakes and loss of time and energy.

The technicians should fix the life for each record depending on the speed of change in technology, fashion cycles, chances of reverting to old systems and discard unnecessary records. Just keeping unwanted documents shall only add to the space and confusions.

Many times, the records are maintained as per instructions of certain senior person, but no review is made. The records are not seen by anybody including the one who instructed to maintain, and the people maintaining the records also do not know as to what purposes this record is maintained. This type of records only increases the work and cost and does not add to the value. It is suggested to periodically review the records that are maintained and rationalize them so as to have the required records and discontinue the unwanted.

The records become redundant and useless because of the changes in technology, which eliminate certain recurring problems by root, introduction of new product and new systems, discontinuing of certain processes, etc. Therefore, periodic review of records maintained is very important, and it is the job of a technical person.

2.22   Reporting the activities

We may do any work, but need to suitably communicate to others so that they shall continue the same. Handing over of charge at shift end is one such activity by which we explain to the next shift man regarding the activities done and pending, the problems encountered and actions taken, instruction given by the superiors and show the actual spots where the work was done. Similarly, we need to explain to our superiors regarding the productions achieved, production lost, quality levels achieved, the problems faced, actions taken, results of the actions, etc. One who can report well is always considered as an efficient person, as he has the facts clear which help in taking suitable action.

The level of understanding and expectations of superiors are different from that of colleagues, and hence, we should be careful and communicate in the same level. We need to understand the expectations and mental status, and suitably explain.

Reporting may be to a person or by written reports. Normally, the shift activities are reported in a register called as log-book or report-book. In order to ensure that no points are missed out, it is better to develop a standard format depending on the requirements. Other reports include production reports, inspection and test reports, maintenance reports, accident and investigation reports, reporting of malpractices or misbehaviours, attendance reports and so on.

2.23   Analysing the reasons

Whatever happened should have happened because of our efforts and not by chance. Hence, it is essential to analyse the reasons for deviations in process and product performance and take suitable actions. The role of a technician is very important in this aspect. He should not have any bias while analysing the facts. He should never come to a conclusion just by his experiences, but should verify the deviations from different angles and take suitable decisions. Management expects a technical supervisor to take logical decisions by analysing the facts and figures so that the company can work smooth.

2.24   Managing the activities in time

'Time will pass and we will pause. Finally we blame time for passing without pausing'.

One of the very important requirements of a supervisor is performing his works in time. Time is a fixed resource. We cannot generate, modify, increase or decrease the time. In reality we cannot manage time. We can only manage our activities to accomplish with in the time frame.

Time management is about getting more value out of our time and using it to improve the quality of life. Time management is a set of principles, practices, skills, tools, and systems working together to help us get more value out of our time with the aim of improving the quality of our life. The time management is not necessarily about getting lots of stuff done, it is making sure that we are working on the right things, the things that truly need to be done. It means to utilize our time for doing right things.

Elimination of unwanted activities is the first step in time management. We are doing number of works without knowing which are really required and which are not. Focus on the goal. Identify the works that are not essential for getting those goals. Group the works as essential and avoidable. Avoid the avoidable works

Be a "miser" – merge, improvise, simplify, eliminate and reduce and avoid wastage of time. Understand the urgency-importance matrix and allot the jobs after sequence analysis wherever applicable. Do the programme valuation and review and adopt lean initiatives. Maintain a time log and monitor the utilization of time for essential activities.

Good Technician can plan and implement irrespective of the circumstances

2.25   At the end

The role of a technician in textile and apparel industry management cannot be just written like what I have written here, as there are still so many things which I am unable to remember and write. I can only say that a technician is the heart of management, and the success of industry depends on the dedicated efforts of planning, coordinating, implementation, review, and actions taken by the people on the spot. We have the habit of blaming others for failures, but it is not correct. If I am a technician, you are also a technician. You can very well apply your mind and add or delete to whatever I say, but have an open thinking, and make this list better.

Policy deployment and middle management

3.1 Introduction

Any organization shall have a purpose and needs to work to achieve the same. The policies are decided by the top management and it is the duty of middle management to implement the same organization wide and demonstrate the same in their routine works.

We can see impressive policies displayed at prominent places and in the advertisements of the organization. It is important to note that a policy written on a paper has no meaning unless implemented, which is needed to be done at all levels and each one should be clear about his role and responsibilities in implementation of the policy. Each one should have targets for accomplishment, that can be measured – key result area. The role of middle management is very important in deploying the policies and achieving the goals.

3.2 Tasks in policy deployment

The top management specifies the mission and policies of the company and decides the goals. The chief of marketing and his team discuss with the customers and study the market situation and advise the top management on what could be sold. The chief of the organization (C.E.O) along with the chief of marketing works together to formulate the marketing goals and plans on a long term basis and give an annual plan. The plant in-charge in consultation with the marketing head works out the product and production plan on a long term basis and for the year. Targets are fixed for each department and section. The strategies are developed to accomplish the plans. The departmental heads workout monthly and weekly plans by involving shop floor supervisors. The supervisors consult the section heads and prepare daily plans considering the day to day positions and the urgent requirements. They need to implement and monitor the activities. In the process, they need to perform number of tasks.

The tasks involve:

- Understanding the works to be done for the day by taking round before the start of the shift and discussing with the previous shift supervisor.

- Verifying the process stocks and the resources availability and arranging for the same.
- Planning and allocating the works for associates depending on the requirements and their competency levels.
- Communicating the customer and work requirement down the line and ensuring that the associates have understood it properly.
- Ensuring that the instructions are clear and the activities are done as needed.
- Monitoring the production and quality by taking periodic rounds and observing the quality being produced and the production taken out at each section.
- Identifying the weak performers and observing their way of performing a job and guiding them to improve performance.
- Identifying the erring performers, training and/or counseling them.
- Ensuring that the nonconforming materials are segregated and kept in designated place and are prevented from getting mixed with regular materials.
- Identifying the machines giving poor quality, and arranging for their rectification.
- Monitoring the wastes generated at each step of the process and taking corrective actions to reduce them.
- Reporting the activities to higher authorities in time and getting instructions incase of doubts if any.
- Maintaining the records for the activities done, production achieved, breakdowns, stoppages, etc., on day to day basis. Machine-wise analysis is needed to identify the poor performance.
- Analyzing the reasons for poor performance like loss in efficiency, low productivity, higher wastes, poor quality and taking corrective actions in time.
- Addressing the grievances of the subordinates and counseling them.
- Representing the grievances of subordinates to the top management without bias.
- Maintaining good house keeping.
- Ensuring that the materials are kept in their designated places.
- Ensuring that the safety gadgets are being used by all concerned.
- Ensuring that the safety gadgets in machines, electric control panels, lifts, etc., are maintained and kept in operational condition.
- Guiding the people to evacuate in case of fires and in emergency situations.
- Arranging for first aid and reporting the accidents as and when happened.
- Maintaining the discipline in the work area.

- Maintaining all the people working under him as one team and taking the leadership.

The supervisor has to plan his activities by prioritizing as per the requirement of the day and ensure that the works are done.

3.3   Steps in deployment of a policy

The first step in deployment of a policy is to have a focus on a shared goal and communicating the goal to all involved in its achievement. It is needed to involve people on the spot and make them accountable for achieving their part of the plan.

The technicians have a main role in understanding the company policies and translating them into departmental and sectional objectives and goals.

One need to identify the links between the company objectives and departmental objectives and prepare action plans to achieve those objectives. These action plans should become the base for writing procedures and work instructions on which the workforce need to be trained. While preparing action plans, the technical aspects of the process needs consideration and the success depend on the technical capabilities of the technicians on job.

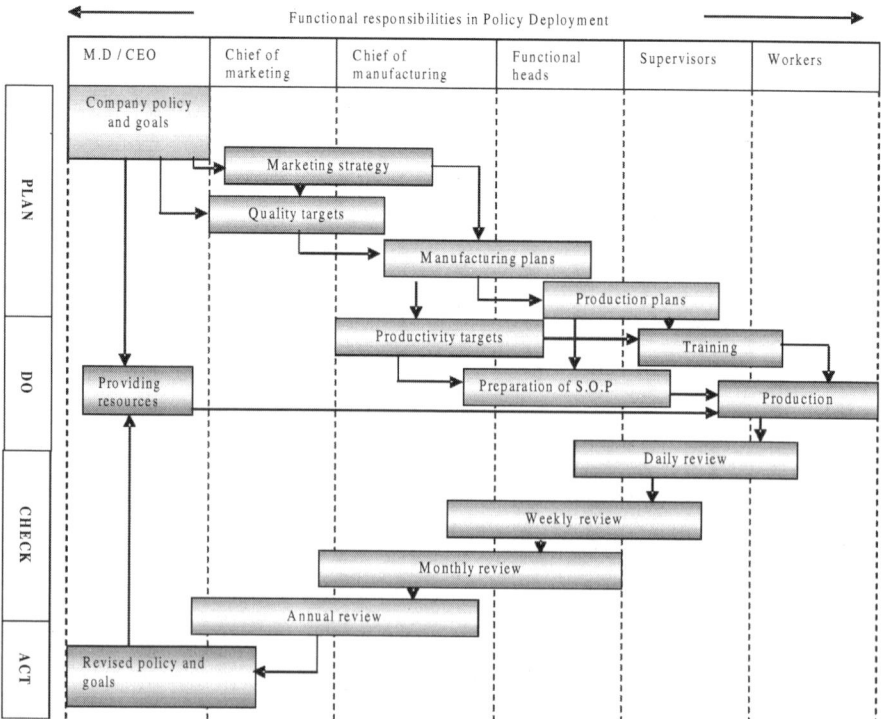

Concern-3	Concern-2	Concern-1	Company Concerns		Departmental Objectives	Objective-1	Objective-2	Objective-3	Objective-4	Objective-5
			✔	Plan - 4					✔	
	✔			Plan - 3			✔			
✔				Plan - 2				✔		✔
			✔	Plan - 1	✔					
				Action Plan						
				Company Quality Policy						
				Company Objectives						
			✔	Objective - 1	✔					
			✔	Objective - 2				✔		
✔				Objective - 3			✔			
	✔			Objective - 4					✔	
	✔			Objective - 5						✔

The technicians need to review their activities and ensure that the goals are being met. If there is any deviation in achievements, the reasons are to be found out and the process needs modification so as to achieve the targets. The activities needs documentation and records are maintained for reference and future actions. The clarity in writing the records and documents is very important so that anyone who refers could be able to understand and implement the actions as needed without any confusion.

Deploying the departmental objectives

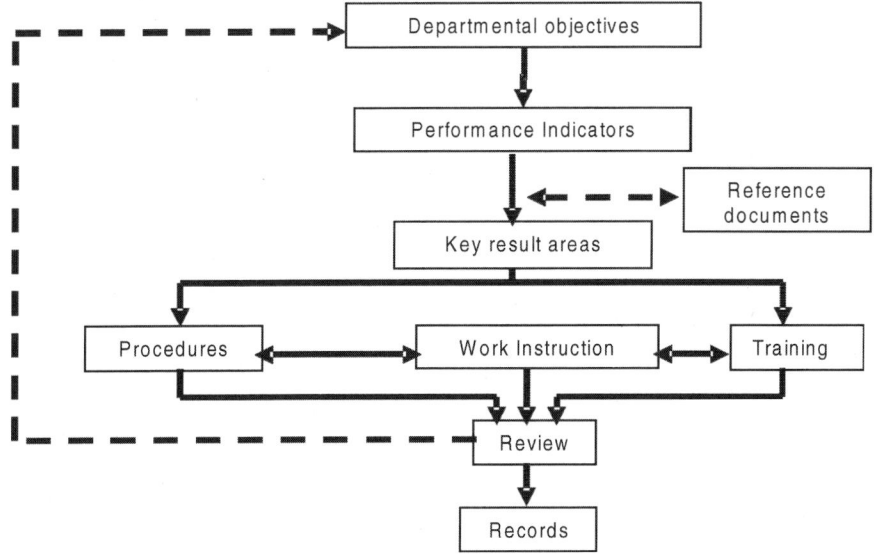

The responsibilities in achieving the results can be broadly specified as follows.

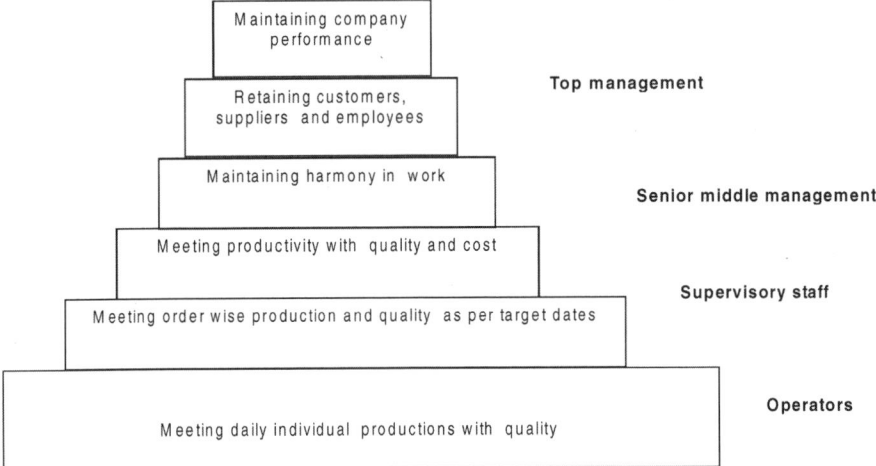

Let us take an example of a training department. The objectives of training department can be written as follows:

- To provide required training to the employees to meet the product requirements and to achieve company objectives while adhering to legal and regulatory requirements.
- To enhance the skill of employees to help them improve productivity and quality while reducing the wastes.
- To build confidence in the employees regarding their ability to perform and to achieve.
- To build the feeling of oneness among the employees.

The performance indicators for the above objectives shall be:

- Reduction in market complaints relating to worker controllable errors.
- Improvement in worker efficiency.
- Reduction in wastes relating to worker errors.
- Clarity about the job content, the responsibility and authorities among the employees.
- Clarity about the product requirement among the employees.
- No loss of production due to shortage of required skills.

The key result areas shall be

- Number of people trained, department wise, subject wise, skill wise, in-house and out-house, on-the-job and in class room.
- Number of training classes conducted, department wise, subject wise,

skill wise, in-house and out-house, on-the-job and in class room.

- Improvement in average skill level among employees.
- The changes in syllabus/curriculum made to enhance the level of learning and knowledge.
- Number of group discussions, experience exchange programme, brainstorming, seminars and conferences conducted and the participants attended.
- The results of training evaluation for all the trainings conducted: in-house as well as out-house.
- No repetition for the training need identified for the same person.

To achieve the results the following procedures need to be specified:

- Planning the training activities and training resources
- Preparing training calendar
- Evaluation of trainers, both in-house and external faculty
- Evaluation of trainees
- Evaluation of training activities
- Preparation and evaluation of training syllabus

The records needed to have proper monitoring are

- The persons trained subject wise, section wise
- The training hours per employee
- Training evaluation report
- Evaluation report of training faculty
- Evaluation report of the training aids, machines, equipments, slides and other presentation materials, books and manuals, slogans and displays etc.

The following reference documents are needed to perform effectively

- Operating instructions
- Standard work practices
- Work norms
- Product safety requirements
- Company standing orders
- Job descriptions

If the supervisors ensure that the activities are done as per the procedures and achieve the sectional objectives of their sections, it shall lead to the achievement of company objectives. Therefore, while specifying the key result areas and targets for individuals, one should ensure that the sum of all results must link to the overall objectives of the company.

Job description

4.1 Introduction

One of the reasons for delays and errors in an activity is not having clarity on the tasks to be accomplished with clear authorities and responsibilities. In order to perform the jobs effectively the tasks are to be specified for each in the organization and without overlapping of authorities and responsibilities. The people on the spot should be clear about their role in the activity and in achieving the company objectives. In number of organizations, the H.R.D. section is given the task of documenting the job descriptions, whereas, in reality the technical staffs working in the shop floor are the correct persons to specify the tasks required to be accomplished to get the results. The H.R.D. personnel can only ensure that the people with competency are recruited and no one is overloaded, but cannot specify the technical activities to be done to accomplish a target.

4.2 Tasks

The tasks are the work one is supposed to carry out. Some tasks are of routine in nature and others are done as and when needed. The tasks, therefore, can be specified as routine tasks and special tasks depending on the nature of work and the type and size of organization. However, some mills do not want to specify anything as routine or special, but would write all as tasks in order to avoid legal complications in case of any dispute.

4.3 Responsibilities

It is essential to specify the responsibility for each, and make them accountable for their works to achieve the targets. The responsibility should be linked to the task. There are some responsibilities like adhering to company's general code of conduct, safety norms and ethics, etc., which need not be specified for each job, but can be displayed as general responsibilities for all.

4.4   Authorities

To perform any task effectively, the concerned person should have the required authority. The authority and responsibility should be balanced and complementary to each other. The authorities should help the performer in preventing the errors and poor quality and also in achieving the efficiency as needed. With inadequate authority, people cannot perform their tasks effectively.

4.5   Minimum competency level

One should be competent enough to accomplish the tasks effectively with responsibility and exercising the authorities. The university degrees and diploma can give a clue as to whether the person knows the basics of the job or not, but cannot be taken as competency. The degrees and diplomas give some confidence in the candidate to handle the task, but unless they are trained and exposed on the job, the competency cannot be developed. The competency required depends on the nature of work, the technology involved, the risks and the skill needed to perform the job. The candidates should have the basic knowledge of the nature of the activity, the works involved, and the probable problems to be encountered so that they can be mentally prepared to take the assignment. Some jobs require maturity for decision making, and in such cases, the experience of handling an assignment helps in judging whether a person is competent or not. It is necessary to specify the minimum competency required for performing a job to help recruiting suitable candidates for the job. The technicians on the spot should observe the works and specify the minimum competency needed for a job. Normally, the minimum competency levels are specified in terms of educational qualifications, experience in certain level, skills, training undergone along with the physical requirements for the job.

4.6   Examples of job descriptions

Let us go through some examples of job descriptions written in textile and garment industries. It does not mean that the job descriptions written here are ideal one, but are examples from the industry. One need to workout the job descriptions depending on the technology adopted, the works that are assigned for each category of people, the culture of the organization in maintaining disciplinary aspects and the level of maturity attained.

4.6.1   Job Descriptions of a floor in-charge in garment manufacturing unit

Following is an example of job description of a floor in-charge in a garment factory.

Tasks

- To understand the production requirements and ensuring allotment of the shop floor workers like tailors and helpers for different batches/styles as per their skills.
- To ensure provision of tools, safety gadgets, and instructions in time, required to the associates (workers and staff).
- To arrange the materials required for batches as per the requirement of the orders/style under production.
- To supervise the activities of the supervisors and the batches in the production floor and ensuring that works are being carried out as planned.
- To monitor the production of each supervisor on hourly basis and arranging for their display on the board.
- To enquire and investigate the reasons for low production, if any, and initiating suitable corrective action.
- To ensure suitable recording of the events and communicating to the superiors.
- To take active part in implementing quality management system in the shop floor as per the documented procedures.
- To coordinate with quality control team in the investigations relating to quality problems.
- To coordinate with maintenance team for conducting maintenance activities as per schedule.
- To coordinate with quality control team for timely inspection of the materials in process.
- To coordinate with maintenance team for investigations related to breakdowns and problems.
- To coordinate with the HRD (human resource development) for the selection and training of associates for specific jobs.
- To participate in the preparation of work instructions, job descriptions, etc., as required.
- To organize for fire fighting, first-aid, shifting of materials and machinery etc., in case of emergencies.
- To do any other odd job as per the situation.

Responsibilities

- Getting instructions from production manager from time to time for the activities being carried out.
- Monitoring the activities of all batches in the floor.
- Achieving production as per standard/requirement.
- Guiding and instructing the supervisors in achieving quality and productivity.
- Ensuring maintenance of discipline in the floor.
- Maintaining house keeping.
- Reporting to production manager regarding the events that took place in the floor.

Authorities

- To discuss with production planning section and production manager in allocating different batches for the styles being produced.
- To report the misdeeds if any of the supervisors.
- To allot/change jobs to subordinates considering their skills.
- To grant leave/permission for the subordinates in the floor
- To stop the production activity in case of any deviations found in the quality till the problem is rectified.

Minimum competency level

- Able to understand the activities, styles, the quality and production requirements.
- Able to read and understand the communications from the top as well as from the subordinates.
- Able to communicate and guide the supervisors and workers
- Mature enough to understand the quality and production problems and take decisions to correct them.

4.6.2   Job descriptions of a supervisor in a garment manufacturing unit

Following is an example of job description of a supervisor in a garment factory:

Tasks

- To get the instructions from floor in-charge and understanding the production requirements in the batch allotted.

- To allot the associates for different jobs in the batch as per their skills.
- To provide tools, safety gadgets and instructions in time required to the associates.
- To get the materials required for batches as per the requirement of the orders/style under production.
- To supervise the activities of the batch allotted and ensuring that works are being carried out as planned.
- To check the production on hourly basis and displaying on the board.
- To investigate the reasons for low production if any and taking suitable corrective action.
- To provide the machines to maintenance team for conducting maintenance activities as per schedule.
- To coordinate with quality control team and getting timely inspection of the materials in process.
- To record the events suitably and communicating to the superiors in time.
- To take active part in implementing quality management system in the batch as per the documented procedures.
- To work together with quality control team in the investigations relating to quality problems.
- To work together with maintenance team for investigations related to breakdowns and problems.
- To participate in the preparation of instructions, job descriptions, etc., as required.
- To take active part in fire fighting, first-aid, shifting of materials and machinery etc., in case of emergency.
- To do any other job as per the situation.

Responsibilities

- Monitoring the activities of the batch.
- Maintaining discipline in the batch.
- Maintaining house keeping.
- Achieving production as per standard/requirement.
- Reporting to floor in-charge regarding the events that took place in the batch.

Authorities

- Authorized to allot or change jobs to associates considering their skills.

- Recommending for leave and/or permission to the subordinates in the batch.
- Stopping the production activity in case of any deviations found in the quality.

Minimum competency level

- Able to understand the activities, styles, the quality and production requirements.
- Able to read and understand the communications from the top as well as from the workers.
- Able to communicate and guide the workers.
- Mature enough to understand the quality and production problems and take directions to correct them.
- Good and sharp eyesight to detect faults while taking rounds.

4.7   General requirements of production supervisors

The production supervisors need to concentrate on getting the production in time of the required quality as per the plans allocated. They need to coordinate with the maintenance supervisors and get their problems solved in the machineries. They also need to coordinate with the quality control section and understand the quality levels being achieved and the problems in the section and take suitable action. The training curricula for supervisors need to concentrate on developing practical competencies in achieving the production and efficiency not only by applying technical knowledge but also by taking all the workmen with him. It should suggest the required basic knowledge to achieve that practical competency. This should have a link with the job descriptions of the supervisors specified by the industry. They should have certain practical competencies and have underpinning knowledge to attain that competency. The common points for all production supervisors in any industry need to include the following:

(a) Practical competencies

- Taking round of the work area before the start of the shift and observing the working. Noting down the important points where action is required or clarification needed.
- Understanding the works done and to be done and taking charge from the previous shift supervisor.
- Understanding the production plan, machines allocated and deciding on further allocations of the machines for different activities.

- Understanding the changes to be done in the forthcoming shift depending on the production plan.
- Verifying the stock of various materials in process and working out the requirement for the shift.
- Checking the quality of materials being produced by taking rounds and informing the concerned maintenance personnel for correcting in case of poor quality.
- Allocating workers on the machines considering the skills and workloads agreed.
- Checking the productions periodically of all machines in the shift and taking action where the production is low.
- Working out the changes to be made for change in product like hank/count/sort/style or product mix and giving instructions to fitters and jobbers and getting them done.
- Monitoring the humidity and temperature as per requirement by coordinating with the concerned engineering operator.
- Checking the conditions of material storage and handling facilities and getting them rectified if found not alright.
- Checking the colour codification and getting them corrected if found deviated.
- Checking the sweeping wastes for preventing good materials going in the wastes. Identifying the areas from where good material is going in waste and tackling the erring persons.
- Counselling habitually absent workers and monitoring their attendance.
- Counselling/influencing a low producing worker to produce as per norms and monitoring their productions.
- Recording the stoppages and working out the production loss. Identifying the causes and working out to reduce the stoppages in future.
- Recording the activities in log book (report book) and submitting to next shift and also to higher authorities as needed.
- Ensuring the use of safety gadgets like caps, masks, gloves and shoes by all.
- Verifying the working of safety stop motions in all the machines and getting them attended if found not alright.
- Maintaining records of production, making summaries as needed and submitting to higher authorities.
- Participating in the mock drill for fire fighting and first aid along with workers.

(b) Under pinning knowledge

- Importance and functions of various machines and mechanisms used in the section in achieving quality and productivity.
- Production balancing – importance and methodology.
- Role of humidity and temperature in maintaining quality and productivity
- Work loads, work allocation and standard working conditions.
- Calculation of production and efficiency and comparing with the industry norms.
- Factors affecting productivity.
- Roles and responsibilities of a supervisor.
- Colour codification and its importance.
- Basic supervisory skills – listening and observing, communicating, counselling, taking charge, reporting and motivating.
- General management knowledge of managing subordinates, coordinating with workshop, electrical department, stores and production.
- Standing orders and discipline in working.
- Precautions to be taken while working.
- Importance of cleanliness and personal safety.
- Fire fighting and first aid.
- Safety gadgets in a textile/garment factory.

4.8   General requirements for maintenance supervisors

The maintenance supervisors have the prime responsibility of maintaining the machines all the time and ensure that they are set properly and give the optimum production and quality. They need to plan the activities in such a way that the productions are not affected. They need to have basic knowledge of the production activities and expert knowledge of the machines they are handling. The common points for all maintenance supervisors need to include the following:

(a) Practical competencies

- Taking round of the work area before the start of the shift and observing the working and noting down the actions to be taken on priority.
- Taking charge from the previous shift supervisor.
    - Noting down the working problems if any in the machines.
    - Understanding the quality complaints in the machines.
    - Understanding the works done till now and the works pending

in the machines stopped for repairs, maintenance works and modifications.

- Understanding the production and maintenance plans and allocating people for different activities.
- Understanding the machine allotted for maintenance and the works to be done.
- Verifying the stock of various parts and lubricants and indenting as needed.
- Referring the machinery catalogue while indenting required spares.
- Checking the quality of spares and other materials received at stores for the maintenance of machines.
- Allocating the workers for different tasks considering their skills and workloads.
- Checking the maintenance activities in the section like routine cleaning, setting, oiling and greasing etc.
- Working out the changes to be made for change in count/sort/style or product mix, and getting the changes made by the concerned maintenance workers.
- Checking the conditions of machine parts while cleaning or overhauling.
- Counselling habitually absenting workers and monitoring their attendance.
- Counselling and influencing a poor performing worker to perform as per norms.
- Monitoring the stoppages due to breakdowns and analysing the reasons.
- Recording the activities in log book (report book) and updating the machine history by entering the spares replaced from time to time.
- Conducting periodic tool audits, i.e., the tools used for maintenance like spanners, gauges, lubricating devices, etc.
- Ensuring proper using of safety gadgets like caps, masks, gloves and shoes by maintenance workers.
- Verifying the safety stop motions and getting them attended
- Participating in the mock drill for fire fighting and first aid along with workers.

(b) Under pinning knowledge

- Importance and functions of various mechanisms used in the machines (automatic and non-automatic machines).
- Planning maintenance activities and replacement of parts as needed.
- Role of humidity and temperature in maintaining quality and productivity.

- Work loads, work allocation and standard working conditions.
- Calculation of maintenance efficiency, and the industry norms.
- Factors affecting maintenance.
- Roles and responsibilities of a maintenance supervisor
- Basic supervisory skills – listening and observing, communicating, counselling, taking charge, reporting and motivating.
- General management knowledge of managing subordinates, coordinating with workshop, electrical department, stores and production.
- Standing orders and discipline in working.
- Precautions to be taken while working.
- Importance of cleanliness and personal safety.
- Fire fighting and first aid.
- Safety gadgets in the factory.

4.9   The control points and the check points for supervisory functions

The supervisors need to know clearly as to what they need to control and for that what are to be checked. The control points are system oriented and the check points are result oriented. The following are the general control points and check points for the supervisors. The specific points for each process are explained in forthcoming chapters.

A. General supervision

Control points	Check points
Production as per schedule	o Speeds, actual Vs programmed.
	o Stoppages and their reasons.
	o Idle spindles/machines and their reasons
	o Breakages/stoppages and efficiency.
	o Supply of back stuff.
	o Production required for completing the running lot.
	o Allocation of machinery for different count/styles/sorts/lots.
Cost control	o Utilization of machines for production.
	o Men employment, productive and non-productive.
	o Waste generation.
	o In process stocks.
	o Avoiding working of non- productive machines/parts.
	o Avoiding good material in wastes.
	o Consumption of stores items.

Quality	o	Quality of input materials.
	o	Quality of material in process.
	o	Quality of materials supplied to next process.
	o	Documentation of activities and changes.
	o	Proper identification of material at all stages.
	o	Handling and storage systems.
Administration	o	Men employment as per norms and designations.
	o	Timings of working.
	o	Discipline in work.
	o	Instructions verified and implemented.
	o	Timely submission of reports.
	o	Clear instruction to next shift / process.

B. Maintenance supervision

Control points

- Procurement of spares, lubricants, tools and gauges of correct quality.
- Use of tools and gauges in proper way.
- Handling and preservation of spares, tools, lubricants, gauges, etc.
- Planning for procurement of spares and maintaining minimum stock.
- Deciding and maintaining the speeds, temperatures, pressures of the machines.
- Engaging trained and skilled persons for maintenance and operating the machines.
- Educating and training the workmen for correct maintenance and work practices.
- Planning maintenance schedules and implementing.
- Allocating works as per work norms and monitoring the quality of maintenance.
- Preparing and providing work instructions and norms for maintenance activities.

Check points

- Quality of spares, lubricants, tools and gauges received.
- Quality of brushes used for machine cleaning
- The condition of the spares removed and the reasons for their wearing out.
- Deviations in machine performance and reasons.
- Alignment of machine parts and setting of various mechanisms and parts.
- Whether there is any part vibrating, which need to be stable?
- Whether the temperature generated by the machines is within control?

- Whether the noise levels are in control?
- Whether the lubricants in running parts are in good condition?
- Whether the time taken for doing the maintenance activities is in control?
- Whether the activities are planned considering minimum loss of production?
- Whether the activ ities are carried out as per schedule?

- Whether the men employed are adequately trained?
- Whether the consumption of spares and stores are in control?
- Whether the loss due to breakdowns and maintenance are within planned limits?
- Whether the quality of materials produced is as per norms?
- Whether the safety gadgets are functioning as per norms?

4.10   Common problems and quality complaints

The responsibility of supervisor also includes attending the quality issues. Therefore, it is essential to explain the commonly occurring quality issues in a factory to the supervisors while they are being trained. In this book an effort is made to list some of the commonly occurring problems, whereas, the individual mill or factory should make a list of such problems they are normally facing and explain to the supervisors during their training.

FACERAP Cards

Fault	Write the name of the fault as it is referred in the industry.
Appearance	Attach a sample or photo of the problem or fault.
Cause	Write the probable cause of getting such fault.
Effect	Explain what happens because of that fault.
Responsibility	Spell out the responsibility to control that fault.
Action	Specify corrective actions in case of observing such fault.
Prevention	Suggest preventive measures to overcome that fault.

While training the supervisors and workers for quality related issues, it is better to prepare a library with FACERAP cards indicating the Fault, Appearance, Cause, Effect, Responsibility, Action and Prevention. These cards shall help the trainees in understanding the problems and also to remember some of the remedial actions in time.

Each complaint is to be taken serious and analysed, and proper preventive actions are to be devised and implemented so that the customers shall get a confidence in the company. The prime responsibility of a supervisor is to ensure that everyone in the section has understood the complaint and the requirements of the customer.

Leadership and self development

5.1   Introduction

There are a number of instances where no other person could do our job, but we are waiting for someone to do without even starting the job. We are ready with excuses for not doing a work; whereas, we are not even make a simple attempt to think on how to do. When someone compels, we do the work. We respect the one who compelled us to do our work, and recognize him as "guru" or "leader". There are numerous examples, like getting up from bed in time, keeping our work area clean, keeping our belongings such as books files records in an organised manner, attending to meetings or training related to our works in time, updating the records in time, documenting our activities for self-assessment, following diet and doing exercises for maintaining self health, planning our activities, having self-discipline, etc. We need to wake up and take leadership to initiate our activities. One should remember that no work can be completed unless it is started, so we should start the work and take leadership for doing our works. Taking proactive leadership in getting the required company's work is the basic work of a supervisor.

A supervisor needs to lead the team of his followers as a leader. To become a leader, one should develop self-confidence and become a role model so that people can follow him. Therefore, leadership and self development go side by side.

5.2   Who is a leader?

Let us understand who is a leader. There are number of definitions.

- One who leads is a leader.
- Leadership establishes unity of purpose and direction of the organization. They should create and maintain the internal environment in which people can become fully involved in achieving the organization's goal. (ISO 9000:2000)
- A person or thing that leads; directing, commanding, or guiding head, as of a group or activity.(Webster's Dictionary)

Leadership is not by birth. One cannot be called as a leader because of the lectures given or articles written by him, the assurances given in public places, the support of goons [unruly people with muscle power] he has, the position he holds in the office, or the strength of wealth with him. One becomes leader by his deeds. Some people might be given high status in the society because they were born in an influential or royal family, but they are not accepted as leaders. People shall throw them out from power when they get an opportunity. People are ready to sacrifice their lives for the leader who looks after them, and never accept the one who is not looking after the interests of public. One who can understand the real needs of people, and work sincerely to fulfill them is a real leader. A supervisor, therefore, should look after the well being of the people working under him, while working for fulfilling the requirements of the company and achieving the objectives and goals. One secret of leadership is that the mind of a leader never turns off.

Leaders even when they are sightseers or spectators, are active; not passive observers. A supervisor should develop this as a culture.

5.3   Why we need a leader?

One can ask a question as to why we need a leader. There are number of reasons. Following reason is widely accepted:

> "To facilitate us to achieve our goals by showing the directions, providing timely feedbacks for us to correct, providing necessary information for enhancing our knowledge and skills, motivating and filling confidence by highlighting our strengths, help avoiding failures by cautioning us in our weak areas."

The leader has certain roles to play and has responsibilities. The normal roles and responsibilities of a leader are as follows:

- Identify and precisely convey the objectives to be achieved to all involved.
- Understand the present situation and workout the strategy to accomplish the goal.
- Educate and guide the people involved by taking them into confidence.
- Follow the activities and provide necessary helps as and when needed.
- Ensure that the goals are achieved to the fullest satisfaction of the people involved.

The word LEADER has 6 letters and each letter stands for one of the important acts of a good leader.

- Listens

- **E**xplains
- **A**ssists
- **D**iscusses
- **E**mpowers
- **R**esponds

A leader should have various properties and some of them are listed as below:

- *Purposeful* – Leading may be for a particular event and need not be for life time. One should be clear about the purpose for which he is initiating and owning the leadership. If the purpose is not clear, he cannot get the correct work done. For example, clearing the materials in the shop floor in the event of a fire accident is different as compared to clearing the materials for despatch to the customer. In the first instance, importance is given to speedy removal of materials from the place and keeping at a safe place, whereas in the second case, cleanliness, proper folding and finishing, proper packing and identification etc., become important and unless they are done properly, the materials are not allowed to move out.
- *Competent* – A leader should be competent enough to lead the team in getting the task done. He should have basic knowledge of the work and also be able to understand the potential of the people associating and distributing the works among them to enable the accomplishment of the task in an effective way.
- *Analytical* – A leader cannot jump for conclusions without analysing the actual situation on the basis of available facts and figures, the clues and symptoms. A good analyst can take proper decision.
- *Decisive* – A leader should take decision and not wait for someone from top to decide. People respect a leader who takes decisions and abides by it. A leader may discuss with others before taking decision but need to take the ownership of the decision on him. For example, the imperfection level of a yarn is in the boarder line, and there is a question as to whether approve it or not. The quality control man suggests rejecting the lot whereas, the production person says that nothing more can be done to reduce imperfections. The marketing people are demanding for the delivery but in the mean time are not interested in inviting a complaint. A good leader takes a decision after analysing the various aspects on a statistical base and arrives at probability of getting a complaint and the expected compensation to be paid in case of a complaint, and the cost and time for preparing the lot again and the losses to the customer for not getting the materials in time.

- *Confident* – Unless the leader is confident, the team cannot be confident and chances of getting results is very low.
- *Appealing* – Whatever the leader says, it should be appealing to others; then only they can get themselves involved whole heartedly.
- *A role model* – When Lord Rams took over as king, his guru Vasishta told him *"Yatha raja tatha praja"*, meaning that the act of people represents the king. People try to follow the foot steps of their leaders as they feel it as the suitable or proper method to come up like their leader. So, whatever the leader does becomes the guidelines for the followers. Therefore, a leader should be a role model and always be very clear in his acts and deeds.
- *Follow the preaching* – People will follow a leader provided he is following his preaching. People follow the deeds of a leader and not the words.
- *Demonstrate* – A leader should be able to demonstrate the way in which the work is to be done; or else, he shall have no rights to demand it from his team.
- *Stand by the results* – All the activities are done to achieve certain results. Leader should have a focus on achieving the results and should be able to change his styles and systems from time to time and ensure the achievement of the result.
- *Shoulder responsibility* – Although, the works are done by followers, the leader is responsible for the success or failure. He should shoulder responsibility and not blame the followers for any failure.
- *Accept defeat/mistake* – A leader should not hesitate to accept a mistake or defeat. One should understand that by accepting defeat and analysing the reasons for failures, one can stand up again and win over the situation.

The characteristics of a leader come through in our day to day interactions with those around us. Leaders come in all shapes, styles, and forms. The Coach4growth.com observes that characteristics of a leader are not skills or behaviors that will be new to those that strive to master them, but will often times be the actions we all know we should be focused on, if we only had the time. Developing good leadership skills take time, just like perfecting an idea or delivering on a project.

5.4   Emotional intelligence (E.Q)

The Santa Clara University and the Tom Peters Group noted that the key characteristics of a good leader normally expected are honesty, competency and forward-looking, inspiring, intelligence, fair-minded, broad-minded, courageous, straightforward and imaginative. It does not mean that these are the only expectations. A number of the characteristics of a leader fall into a greater category that many of the leading executives of today refer to as "emotional intelligence". Achieving this level of leadership will inspire those around and lead teams to great heights. Emotional intelligence or EQ is the demonstration of good leadership skills by leaders who are self aware, self regulating, motivated, empathetic and possess great social skills.

Leaders with a high degree of emotional intelligence do right thing, come to the table with the right technical and knowledge based skills. These leaders inspire and drive those around them. Good leaders do not let their emotion and mood to affect others as they have the ability to control their emotions and to think before acting. They are not easily flustered and are comfortable with ambiguity and change. While leaders demonstrating emotional intelligence are persistent in the achievement of goals, they do not ignore the emotions of others. They strive to understand the emotional makeup of others, value diversity and take that into consideration during interactions. This type of leader is proficient at building and managing relationships, as well as establishing common ground. They are able to influence others, lead through change, and build strong teams.

Leader does facilitate his followers to do their works effectively rather than dictating them and getting the results. Leader as a Facilitator:

- Guides without directing
- Brings change without disruption
- Helps people to self discover new approaches
- Breaks barriers between people
- Preserves the group values

The effectiveness of a leader depends not only on his competency, but also on the attitude and systems in the people or the organization. But the efforts of the leader are more important to change the attitude of the organization or the people around him. People vote the leader on whom they have confidence. One can enjoy as far as he has majority, but when people are not happy, he loses his majority, and also the status. Hence, the leader should be careful and dedicated to the cause of his followers. He should win the hearts of the people.

5.5   Leadership styles

A leader cannot always act in a single style. Depending on the situations he needs to change his styles. There are four styles of leadership. They are as follows:

- Delegating
- Supporting
- Coaching
- Directing

The delegating can be done when the leader is confident about the capability and integrity of his assistants. The supporting style is used when the people working are competent, sincerely making efforts to achieve the results, but are not getting the results due to various factors beyond their control. Coaching is done to build competency in the followers. Directing is adopted where chances of not adhering to systems is found; the reasons may be any.

There are different types of leadership, and depending upon the situation, a leader has to choose the appropriate type. They are as follows:

- Pulling type
- Pushing type
- Leading from front
- Leading from back
- Leading from within

Pulling type is adopted when the subordinates are failing in following the systems or in understanding the systems. The pushing style is used when the subordinates have competency but are not confident. Leading from front is done when the supporters are afraid of a situation or not aware of it; leader stands in the front and shows the way. Leading from back is done when the followers are making efforts in correct path, but one needs to ensure that they do not retard or return back. Leading from within is done in team approaches for various projects.

5.5.1   Beauty of a garland

Each one has some weaknesses, and also some strength. We should know how to encash the strength and impress the people by our strong points. We should also work to overcome our weaknesses. This can be done by encouraging our fellows to improve and shine because of their strengths. Let us take the example of a garland, where the beauty of it is in the placement of flowers in their respective places (Fig. 5.1). Who does this? The thread is the leader, who holds the fellow flowers in their respective places, and shows their beauty to all. But remember, the thread never comes out to show its face as the leader. If the thread comes out, the garland has no value. If leader tries to dominate and not allow his followers to expose, that leader shall be ineffective and the team fails. The thread in a garland is the best example for leading from within.

5.1 Beauty of garland is in placing of its flowers by the strength of the leader, who is holding them together.

People follow the leader who is worth believing. The one who implements his preachings is called an *Acharya,* and people respect them. People shall have faith on him as he does not lie or cheat others. A leader should therefore be an *Acharya* or a role model.

A real leader never hesitates to take risk to ensure the safety of the followers and to motivate them. There are incidences of a leader jumping in front of violent mob in case of violent or unruly incidents in the mills and calming them by virtue of his good leadership. Such leaders are not afraid to the threats given to them.

Good leader ensures that the requirements are met of the followers. A good supervisor represents the genuine cases of his workmen to the management and ensures that their grievances are addressed.

Caption ensures that others are safe before he jumps for safety. For example, if there is a fire in a cotton mill, the leader ensures that untrained and lady workers are evacuated first and he shall be standing inside along with trained people for fighting the fire. He will come out only when the fire is completely brought into control and extinguished.

Leader can show path even where others cannot reach. Leader, by virtue of his experience and foreseeing, shall be able to find a solution to the problem. We can prosper by good leadership

5.6   Supervisor as a leader

Supervisors play a crucial role in daily management. They are the people involved in the design and establishment of the process, coordinating with the people on the shop floor and make them understand, implement the process, monitor and correct them.

Supervisor leads the team of people working under him. Supervisor educates the management on day to day happenings by providing the facts and figures that lead to take decisions. Hence, supervisor also leads the management. [When information given is incomplete or not real, the term 'misleading' is used]

The supervisors are always under the pressure of producing the required quantity in time as per the agreed quality irrespective of the odd situations they face due to various factors relating to inadequate communications, labour shortage or lack of cohesion, power shortage, maintenance lapses, and non availability of critical parts, changes in climatic conditions and sudden changes in customer requirements and so on.

5.6.1   Why supervisors are respected?

The supervisors have no right to give an excuse, but to give results under all circumstances by foreseeing the problems in advance and taking precautionary measures. They have been respected for this ability through out the world. Unless one applies his mind and work out solutions for various problems he face, it shall not be possible for him to be successful as a supervisor.

5.6.2   Supervisor and management

Supervisor is termed as middle management personnel. He is the manager for his section and the shift, as his workers shall obey only his orders. Supervisor is the leader of the section he is looking after. He represents the grievances of his people to top management.

5.7   How to become a good leader?

People say that good leadership results from the individual's attitudes and actions. This is partly correct. Don Blohowiak observes that great leadership results from both individual efforts and the collective support systems in the leader's environment. Good organizational support systems can help a good leader more easily become a great leader. Figure 5.2 given by Don Blohowiak shows how the combination of individual competency and the organizational leadership results in developing great leaders.

Supervisors are a part of organization's management system and need to develop leaders among their colleagues and subordinates, so that they can delegate some of their authorities and responsibilities and have some time left for planning and development.

5.8   Self awareness and development

Self-awareness is about recognizing and understanding one's emotions,

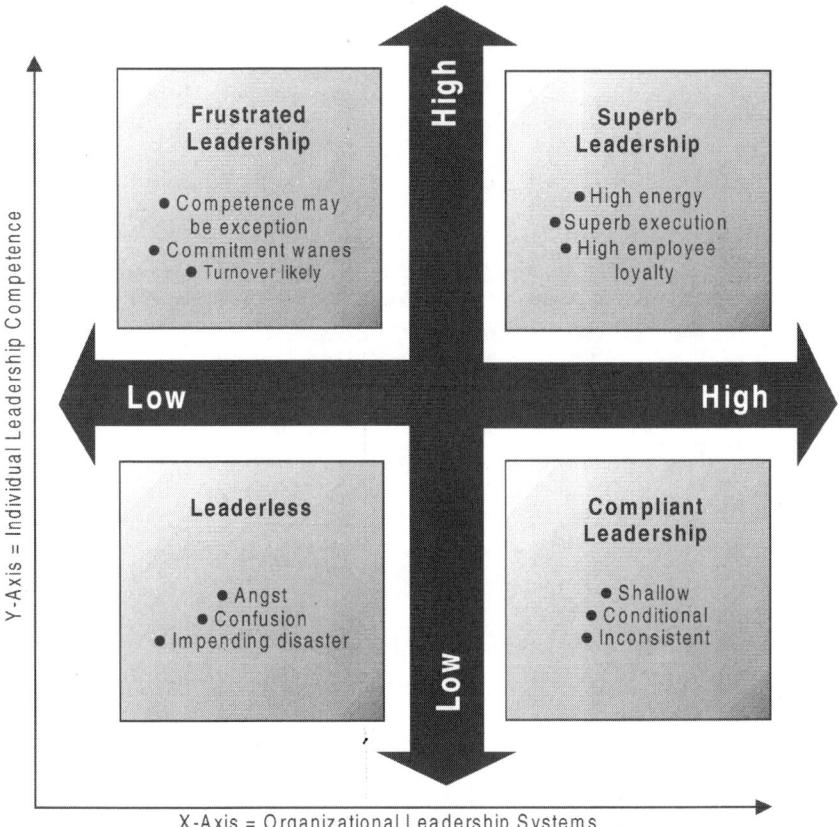

5.2 Leadership system and competence.

moods, and values. These aspects have great influence on one's success as a leader. A self-aware leader is confidant and realistic about his/her strengths and opportunities and uses this to inspire those around him.

Self development is very important for any individual, and especially for a supervisor who needs to be followed by others. In number of cases, inspite of one's efforts to become a good leader, he might be failing due to various reasons like corporate policy or culture, political influences, etc. Don Blohowiak suggests following questions for self review.

- How can I modify what I'm doing to more effectively counter the opposing forces?
- What can I do to eliminate the very existence of such forces?
- Given current trends, what's the likelihood of this situation improving in the near future?
- How long can I remain committed, motivated, and effective in this environment?

- What are my alternatives?

All leadership is contextual and the effectiveness is as much tied to time, place, and other managers, as your own resolve and actions.

5.8.1   Understand self

While planning to develop self, the first question one should ask is "Can I describe myself". Most of the people cannot describe themselves. Unless one understands self, he cannot plan and improve. Normally, people tell the following as their description which are formal:

- Name
- Family background
- Date of birth
- Qualifications
- Jobs done
- Hobbies
- Wealth
- Awards won
- Physical appearance – height, weight, etc

If we really want to understand self, we need to go deep and ask ourselves further questions like:

- What are my ambitions? Why those ambitions?
- What are my normal likings? Why I like them?
- What I do not like? Why I do not like?
- Who or what is my favourite? Why it is my favourite?
- Who are my friends? Why they are my friends?
- Whom I do not like? Why I do not like them?
- Which is my favourite food? Why it became my favourite?
- Which is my favourite taste?
- Who are my competitors? Why they are my competitors? How they compete?
- What is my aptitude?
- What are my attitudes?
- What are my habits? Which of my habits are good and which are bad?
- Am I hasty in taking decisions or cool in thinking?
- Whether I stick to words or change decisions?
- Whether I am lazy or active?
- Whether I am strict or lenient?
- Whether I am conservative or spend thrift?

- How is my relation with my family members?
- Whether I like to remain alone or mix with people?
- What Skills I have?
- Which work I can do better?
- What is my physical ability?
- What are my assets?
- What are my liabilities?
- What is my culture?
- In which area I can achieve both efficiency and effectiveness?

5.8.2   Johari window

Even after asking so many questions and trying to find out honest answers, we may not be able to completely understand ourselves. Famous psychologist Johari divides the knowledge about a person into four groups, which is popularly known as johari window (Fig. 5.3). There are certain points about self which are known to me as well as to people around me, which is shown in first window stating that 'I know and you know'. There are certain points that are known only to me and not to others, like my ideas, my secrets, etc, which are shown in second window 'I know and you do not know'. There are certain things, which I do not know and others may be having full knowledge about me, like my real capability whereas, I am living in an illusion. These are expressed in the third window stating 'I do not know and you know'. There are certain aspects which none of us know as there was no opportunity to exhibit them. For example, in case of an emergency situation like accidents or fires, one might take leadership and bring the situation in control; whereas in normal cases, no one would

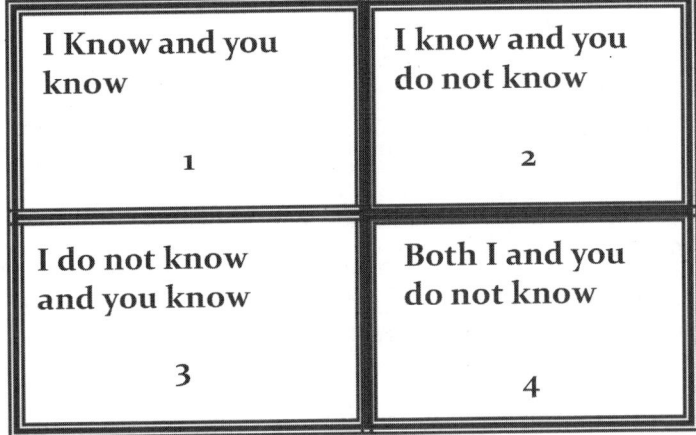

5.3 Johari window.

have expected him to have that capability. This is explained in fourth window, 'both I and you do not know'. If one has to improve, should try to understand about self from the others point of view also by suitably interacting. One cannot simply assume whatever he knows about him is the final.

5.8.3   Achieved and unexplored areas

People spend their time in enjoying their achievements or worrying about the failures. It is important to understand what is achieved and not achieved, and work for achieving in the unexplored area (Fig. 5.4). If one realize, the unexplored area is always more compared to the achieved area considering the real competency one is having.

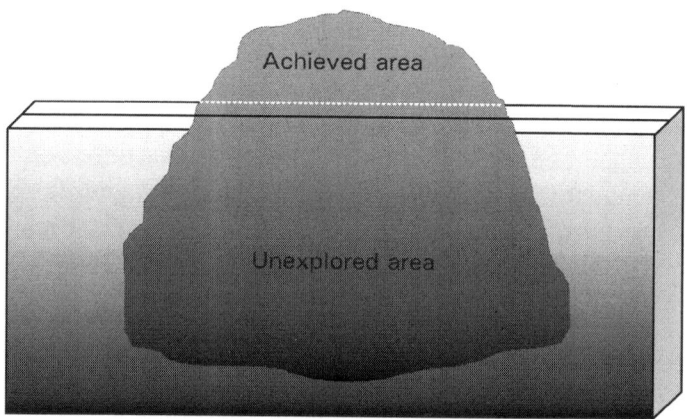

5.4 Achieved and unexplored area.

5.8.4   Nothing is useless

In number of cases we hear that 'people are useless', 'our people cannot do this', 'we do not get suitable persons,' etc. We need to ask ourselves whether we know our people well.

Everyone has potential to achieve certain things, but is not exposed to those situations, and hence remain unnoticed. Have we really made efforts to identify and encourage hidden talent or competencies? There is a Subhashita is Sanskrit stating that

> *"Amantram aksharam naasti, Naasti moolam anoushadam,*
> *Ayogyam purusham naasti, Yojakastada durlabhah"*

It means there is no letter without meaning or that cannot be used for praying, no plant is without medicine values and no person is without use, but there is scarcity of those who can analyze, organize and use them properly. It is true even today. We are not able to understand the capabilities of our people and utilize in a proper way.

5.8.5   S.W.O.T analysis

Each one has certain strengths and certain weakness. It is necessary for one to understand the strength he has and take maximum benefit out of it. By understanding the weaknesses, one can avoid taking risk in such area, where he is weak. He can make efforts to minimize those weaknesses.

S.W.O.T analysis (Fig. 5.5) is identifying and analysing the strengths and weaknesses of a person by self and foreseeing the opportunities and threats. After making S.W.O.T analysis, one needs to work out actions to take the advantage of the opportunities and protect self from the threats.

STRENGTH What are my strengths? What I have achieved so far? What can I achieve further? Where I can be successful?	**WEAKNESS** Where did I fail? Where I am likely to fail? What should be avoided?
OPPORTUNITIES What opportunities I have? How to cash the situation? What should be my target?	**THREATS** What are the threats I have? How to prevent it? How to minimize it?

5.5 S.W.O.T analysis.

5.8.6   Old roots and new leaves

> *"Hosa chiguru hale beru koodiralu mara sobagu*
> *Hosa yukti hale tatwadodagoode dharma.*
> *Rishivaakyadodane vignana kale melavise*
> *Jasavu janajeevanake Mankuthimma."*

The meaning of the above Kannada poem is "A tree is strong with its deep roots and beautiful with new leaves. A combination of strong deep old roots and new leaves are the requirements of a beautiful and strong tree. Similarly, the deep rooted core values and ethics suggested by old rishis in the culture along with new ideas coming as a part of developing science and technology can bring prosperity to mankind". The above is poem from the book Mankuthimmana Kagga, by late D.V. Gundappa.

While doing your work keep the objectives as the base and develop your systems by your imaginations and logics (Fig. 5.6). Be enthusiastic but do not leave the base trying to fly in the air. When you think of a higher productivity, keep the customer quality requirements as the base and develop methods to achieve higher productivity while fulfilling the customer's quality requirements.

5.6 The beauty of a tree is in the combination of its deep old roots and the fresh leaves covering the top.

5.8.7   Enjoy the work

The quality and efficiency can improve when one enjoys his work. One without interest feels the work as boring and gets tired fast although he is physically fit; whereas a person with interest does not get tired by that work. Hence, one should always ask himself

- Am I enjoying my work?
- Do I like my work?
- Which part of my work is good?
- Why it is a good work?
- Which is not a good work?
- Why it is not a good work?
- What I expect out of my work?
- What your management expects out of my work?

People may be having lot of ambitions and dreams, but should not forget that they are already in a job, and that job is providing them food and shelter. They should work sincerely and excel in their job. That can be done only when they love their job. Great philosopher of twelfth century Sri Basavanna (Kannada) says *"Illi salladavaru alliyoo sallaru"* which means 'if one cannot survive here he cannot survive there also'. We need to learn how to succeed and survive in our present circumstances rather than dreaming of a success in another world.

Once one gets married, it is the responsibility of both husband and wife to love each other, adjust themselves to the feelings and requirements of each other so that they can live happily. The same is true when we join an organization and accept certain role. We need to love our job and try to excel, by which we develop potential for excelling in any job or responsibility we take.

5.8.8   Concept of 3-H and 4-H

As the industrialization took place, numbers of craftsmen became jobless and were employed as workers in big industries. Earlier, the craftsmen were doing all the work by their own hands, whereas now the works are being done by machines. It became a practice in the manufacturing industry to refer the workers as hands instead of referring as human beings. The terms "number of hands employed", "shortage of hands" became very common in industry. However, as the quality and productivity started gathering more importance, it was felt that the people working should also have to use their heads rather than using only hands. The managements started insisting on using 2H instead of one H, i.e., hands and head instead of only hands. As the concepts of human approach to management was developed, people realized that the work can be done more effectively when a person likes and enjoys the job, and works with his heart along with using hands and head. This is termed as 3H concept

- 1 H – HAND
- 2 H – HAND and HEAD
- 3 H – HAND, HEAD and HEART

Producing the goods and services to fulfill the customers' needs and expectations is normally insisted to make the company competitive. Whereas, one should understand that respecting human values and ethics are also very important for getting the support from the society for the sustenance of our business. We are a part of society consisting of human beings and we need to respect the human values. All our employees, suppliers, customers and other stake holders are all human beings. So adopt **4 – H** to become successful all the time, i.e., hand, head, heart and human values.

5.8.9   Work culture

Normally, people use the term work culture, which consists of doing the work in a prescribed manner all the time without anybody insisting. Where the work is developed as a culture, it is done regularly without any violations or omissions. Good work culture leads to quality products and also for higher efficiency. Culture is not developed in a day, but over a long period. When people see benefit in any work or system, they try to follow it religiously. Therefore, to develop a work as a culture, the people working should accept the system and realize it as beneficial for them in the long run.

Man, by nature, is lazy when he does not feel or convinced of the necessity of doing a job. He tries to avoid the works whenever a chance is given. But man is not always lazy. Man does not like to sit idle. He wants to do something by which he gets happiness. He enjoys the work done by him when he gets the results he wanted or lives the job he is doing.

An artist or a scientist may work day and night without rest or even food to make their work complete. They enjoy their work and they are happier when they get success in the work and not by any other means. The same thing does not happen with others who are working for someone else. But, if a man is proud of his job, then he becomes dedicated and works for the accomplishment of the job. A good work culture always concentrates on making people proud of their job.

Human beings are supposed to do their work. To work is the culture of humans. Prof S. D. Mahajan in the book "Saad Deti Yasha Shikhare", written in Marathi explains the culture of doing work as follows.

> "Kaam talana hi vikruti aahe
> Kaam karana hi prakruti aahe
> Kaamaatoon ananda milavana hi samskruti aahe"

"Escaping from work is awkward or disability (Vikruti), doing the work is nature (Prakruti) and getting happiness in work is real culture (Samskruti)".

5.8.10   Stress at work

People talk of stress at their work. Some people feel that there is too much stress, especially in the jobs of middle management level, whereas some are enjoying the same jobs. Let us understand the following facts:

- When you do not like a work you feel tired and boring.
- When you have no confidence of success, you feel exhausted.
- If job is not challenging you get bored; one will be less happy in the

victory against a weak challenger than a defeat against a world champion.

When one likes a job, is confident of doing it successfully and ready to take challenges does not feel stress at job; but when the job is forced on one who does not have liking for the job or do not have required competency for performing that job he gets stressed and cannot perform the job as needed.

When people love their job, they work more than what was expected. They do not feel tired or hungry even in spite of working long hours in an unfavourable working condition. They reach to great heights.

5.8.11 Are you disciplined?

People like challenging jobs rather than routine jobs; this is true even for the middle management. However, to get success in the job, whether it is challenging or not, discipline is very important. One even with talent might not get success by the lack of discipline. Discipline is that refining fire by which talent becomes ability. A supervisor should be disciplined so that he can make his fellow workers disciplined and maintain discipline in the work place. Normal disciplines expected by supervisors are starting the work in time, reporting the activities in time, use of soft cultured language while tackling subordinates for errors or mistakes, using safety gadgets while on work, dressing properly as specified for the job, maintaining good house keeping all the time, not by passing any steps in the procedures while performing any work, not violating the instructions from the superiors, and so on.

The supervisor who is disciplined can expect his fellow workers to be disciplined. Therefore, maintaining self discipline is very important for a supervisor.

5.8.12 Are you confident?

In number of cases, it is seen that the people are depressed when targets are not achieved. One should always understand that results are not only because of one factor but a combination of various factors. First analyze whether you are competent to achieve the goal if the conditions are favourable. Then work out whether the target is achievable in the present circumstances. Then you will have confidence. There are number of reasons for not achieving a goal, and do not think that the goal is never achievable. The circumstance might not be favourable or the competency might have to be improved by proper education, training and practice. Once the confidence is developed, man puts extra efforts and can achieve the results.

If a person is not confident, he does not put the required efforts to achieve the results. Therefore, for a supervisor to become success, he should be confident in his job and also should be able to fill confidence among his fellow workers that they will be able to achieve the goals.

5.8.13 Win your internal enemies first

If one has to win over others, he need to first win over his internal enemies first. There are six internal enemies for a human being, termed as "*Ari Shadvarga*" in Sanskrit. '*Ari*' means enemy, '*Shad*' means six and '*Varga*' means group. They are

- *Kama* [Passion, especially for sex]
- *Krodha* [Anger]
- *Lobha* [Greediness]
- *Moha* [Blind affection/Illusion]
- *Mada* [Haughty]
- *Matsarya* [Jealousy]

The above six enemies are the main reasons for the down fall of a person. It is very difficult to win over them and have complete control, but one needs to practice. A supervisor can be successful when he tries and take control over the above six enemies.

5.8.14 Some tips for daily life

Following are some of the points those need to be used in daily activities so that one can come up in life.

- Talk softly
- Walk humbly
- Eat sensibly
- Breathe deeply
- Sleep sufficiently
- Dress smartly
- Act fearlessly
- Work patiently
- Think truthfully
- Believe correctly
- Behave decently
- Learn practically
- Plan orderly
- Earn honestly
- Save regularly

- Spend intelligently
- Love passionately
- Enjoy completely

5.8.15 Ant philosophy

Ant is a very small creature, but there are number of points to learn from it.

- Ants never quit; they move in the decided direction and reach the destination inspite any odds and hurdles.
- Ants think winter all summer; they do not take rest but shall be continuously working in collecting and preserving food for winter and rainy seasons.
- Ants are disciplined. They follow their line without deviating.
- Team work is propagated by ants. They can carry and move a big cockroach by their team efforts.
- Ants always think positive; a small ant carries material which is very much heavier than the weight of that ant. If one ant cannot carry, it takes the help of other ants and carries the material to the destination crossing all hurdles.
- Ants have no limit. They are ready to go to any distance for collecting their food.

5.8.16 Be mad but not a stupid

Huchcha neenaagu pechchanaagali beda
Kechchedeya kaliyaagi uchchaatisnhateya
Nechchadiru sihinudiya mukhastutiya maatugala
Vechchamaadali beda samara hechchilla.
Nechchi manadali Purushothama Vittalana naama
Echcharadi munnadeyo nee Huchchuraama.

The above poem is taken from the Kannada book "Huchchuraamana Muktakagalu", a guide for daily management. It says, be mad but not a stupid. A mad will not leave the path or the work what he has decided. He shall be always after the same thing, whereas a stupid does not know what to do and which path to select. Become a knight by overcoming your weaknesses. Do not succumb to the sweet praising that is normally with an intention of getting favour from you. Do not waste time as it is scarce. Keep faith in god, but be alert while moving forward.

Individuals and teams

6.1   Need for building team

One may claim that he comes alone while coming to this earth and also goes alone while going; but he cannot stay alone. Even when we analyze, he did not come alone; there was effort from minimum two persons to bring him on this earth. While he goes out also four people are needed to take him out. Hence man is always depending on others.

One might claim that he did or he achieved something single handed; but it is not correct. For all practical purposes, one can accept that he achieved that task alone; like a scientist developing a new concept, a swimmer swimming and winning gold in a competition, an athlete running and winning gold, a worker producing record production in a given time and so on. If we carefully analyze, we can see efforts of number of people in making him success in achieving that task. It might be the one, who is providing him food, or the one washing his clothes, or the one who did not create any disturbance in his work so that he could succeed, or the one filling enthusiasm by encouraging in all his acts, or the one who guided him during his works. Man is a social animal and has to live with others like relatives, friends, colleagues, society and so on. So we have to accept that a man needs help from others to make his task complete.

It is essential for anyone to work in teams. A supervisor in textile or apparel industry also has to work in teams. He needs to work in different teams; may be as a member in the team of fellow supervisors, as a leader in the team of fellow workers, as a member of management team or various other teams. Building teams, maintaining harmony in the team and working for success of the team is one of the important tasks of a supervisor.

In textile and apparel industries, certain works need to be done in groups only, whereas, some are allotted to individuals; however, they also have a relation with some other activities that are done. The group activities like doffing a speed frame, doffing a ring frame, cleaning and maintenance activities, beam gaiting on looms, number of wet processing activities, material handling, stitching activities in garment manufacturing etc., demands synchronization of works between the team members and cooperation between members. The supervisor has to ensure proper coordination between the activities of team members and also harmony in working. The success of a supervisor depends more on his skills in coordinating between the team members rather than his technical expertise in setting a machine for various parameters.

The teams may be for achieving a fixed task or to work continuously in harmony for a long time. The short term teams are normally formed to achieve certain tasks like preparing a curricula for certain training, erecting and commissioning a new machine, analysing a complaint and suggesting corrective and preventive measures, taking improving projects for reducing wastes or improving quality or improving productivity, and so on. There are certain teams that are having a long term goal of working in a harmony and cooperation among members in monitoring certain activities like safety, welfare, and ensuring compliance to certain requirements of trade and social compliances. The examples of such long term committees are quality circles, work committees and joint committees.

6.2   Formal and informal teams

The teams may be formal or informal teams. The formal teams are formed as per certain norms of the company or the government. The examples of formal teams are joint committees, work committees, welfare committees and safety committees etc. They function as per the guidelines given by either the government or by trade. The activities involve verifying the adherence to safety norms, human rights, grievance redress, welfare measures, canteen management, welfare activities, and social compliance and so on. There are certain formal teams, normally referred as cross functional teams, where members from different sections are nominated by the top management to accomplish certain tasks within a certain time period. It may be installing a new plant, shifting machines from one site to another, attacking certain chronic problems, preparing curricula for training modules, conducting sales promotion programme or customer meets.

The informal teams are formed either for performing certain tasks or with an intention that people in a team work together as one all the time. The examples are certain committees formed for arranging sports activities, cultural activities and quality circles.

A team is more than just a work group; a collection of individuals who work together. In a work group, each member is directed by and reports to a common manager or supervisor, but members don't necessarily collaborate with each other to complete their tasks. The manager of a work group usually has deciding authority. A team, by contrast, comprises individuals with complementary skills committed to a shared purpose, common performance goals, and an approach for which they hold themselves collectively accountable. The members interact with each other and with the team leader to achieve their shared goal. In a team, members depend on one another's input to perform their own work and look to their leader to identify and provide needed resources, coaching, and a connection to the rest of the organization. A team makes decisions that reflect the know-how and expertise of many people, not just the leader. Figure 6.1 depicts the nature of a team.

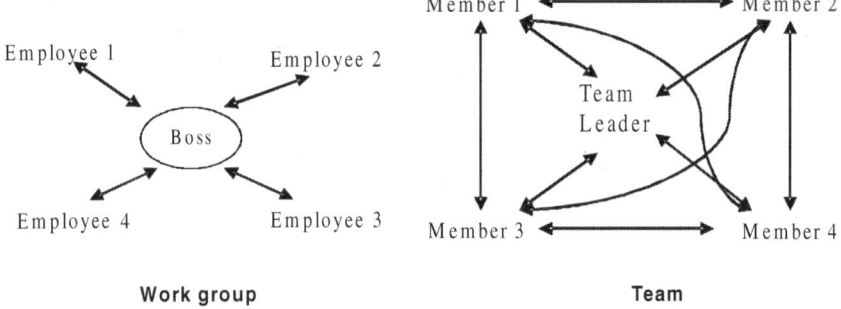

6.1 Work group and team.

The doffing gang in a spinning mill or a batch of tailors in a garment factory belongs to the category of group; whereas a mounting team in carding belongs to the category of team. The doffers or individual tailors in a batch are not normally interacting with other group members for their works whereas, a team member in mounting of a card has to continuously interact with other team members. Although, there is interaction between members in a team, there is a fixed authority and responsibility with clear job contents for each member. Each one has to perform his or her job in time as per the schedule to make the team effective. The supervisor has a task of ensuring that all are working as per their assigned jobs.

6.2.1   Different teams, different purposes

Organizations form different types of teams for different purposes. Here are some examples:

Team type	Purpose	Example
Self-directed work team	Meets are ongoing, and may be on daily basis to perform a whole work process	At a cotton spinning mill, a team of four people is responsible for ensuring that raw materials are purchased correctly according to company guidelines.

Project team	Gathers to address a specific problem or opportunity and then disbands. They need not meet daily, but as decided by the team.	1) Several unit heads in a garment factory explore the potential benefits of adopting a new technology, present their findings to executives, and disband. 2) A team taking the assignment of installing a boiler and commissioning.
Virtual team	Brings geographically separate individuals together around specific tasks	A project manager for garment brand hires agents from around the world to work with garment factories on a major style; while making all transactions on internet and ERP.
Quality circle	Works on specific quality, productivity, and service problems	1) Customer-service employees and managers in a process house generate and implement ideas for improving service to the company's biggest customers. 2) A team of loaders work together and design a new method of unloading and loading.

6.3   Characteristics of an effective team

Whatever may the type of team or the task given to them, the team should perform. The word PERFORM has following letters, which are the characteristics of a good team.

- **Purpose**
- **Empowerment**
- **Relationship and communication**
- **Flexibility**
- **Optimal productivity**
- **Recognition and appreciation**
- **Morale**

It means the team should perform; if it is not performing, it is not a team, but just a mob.

6.3.1   Purpose

Everyone has a purpose and works to fulfill it. The organization has a purpose and in the same way any team has a purpose. The team leader needs to ensure that purposes are fulfilled. Therefore, the leader and the members should be clear about the purpose. To achieve the purpose, the activity of the team shall be decided and each one has to be given certain objectives and goals. The leader should align company objectives and goals with individual objectives and goals. The leader should convince the team members that by achieving company/team goals, one can help achieving personal goals.

In the process of understanding the purpose following steps may be considered:

- Enquire your boss why you should do this work.
- Understand why you are doing this work.
- Explain to your people why they need to do this work.
- Understand from them why they are doing this work.

6.3.2   Empowerment

Anyone can work provided he is empowered suitably to perform the task assigned. They should not hesitate while doing a work with a confusion of whether to do it or not, whether I shall be questioned if I do the job and so on. Therefore, empowering the people to do their jobs with delegating adequate authorities to perform the jobs is very important. Please remember, as a team leader, if you do not empower your people, then you have to do all their jobs. When people are empowered, they shall do the work with confidence and interest that help in improving efficiency and productivity. Mind works faster when there is no tension, and efficiency increases when there is no force.

6.3.3   Relationship and communication

There is no meaning by staying together without good relation. When there is a good cohesive relation between team members, they work together. One has to remember the saying "together we win; divided we lose".

Trust is the driving force for relationship. The leader should:

- Have trust on the members
- Understand the feelings of the members
- Communicate with members effectively
- Work together with the team members
- Be flexible; do not be rigid all the time. The rigidity may be detrimental to relationship.
- Concentrate on results

Mode of communication and the methodology used has a say on the relationship between members of a team. Written communication gives clarity to the members, but should be used only for certain items where remembering is difficult. For example, when a trail is decided, the process parameters may be given in writing. While trying to understand certain act of a member, a written communication shall lead to displeasure, whereas, an informal talk shall be beneficial. Some time phone calls may not be helpful as the members might be busy in certain activities and not able to take the phone, and in such cases a SMS message or a slip of paper indicating the message in simple words may help. Sometimes, the written message might not be conveying the correct intention of the message and might lead to misunderstanding, in such cases, it is better to talk personally and convey the correct message.

6.3.4   Flexibility

The important point in a team is that it should perform as one, and no individual should be given undue weightage. The members have to observe certain code of conduct and the leader should be rigid in enforcing the discipline and ensuring that the code of conduct is respected. However, while working in a team comprising of people with different ideologies, if it has to perform as one, some rigidity may have to be eliminated, but the focus on achieving the goal shall remain intact. The leader should decide on how much he should be flexible with his members and in the procedures to be followed so that there is no negative feeling among the members and they work as one to achieve the objectives.

6.3.5   Optimal productivity

Achieving optimal productivity is one of the tasks of any team, and it should focus on the same. What is required by the team and for the company is to be understood by the team members and the leader. They should be clear as to how much is required to be produced and has to work for achieving the results. The leader along with the team members should review and find the reasons for non achievement. The leader should facilitate the team members to achieve optimal productivity.

6.3.6   Recognition and appreciation

Recognition improves efficiency and brings cohesion in the team. If you do not recognize in time, people get de-motivated. The interest in the team as well as in the work reduces when recognition is not given for their efforts. In a number of cases, the management recognizes the people only when they get the results, whereas the results are a function of number of external factors. Let us take the example of a marketing team. Due to adverse market condition, the mills might not get required price for yarn although the marketing team has done good efforts. Alternately, the quality of the yarn may be bad and because of that the marketing man is not able to push that yarn in the market. The marketing might have been successful provided the members had freedom to negotiate the price, whereas, the price is fixed by top and the marketing cannot book the order with lesser price even by 1%. Therefore, one has to see the process and recognise the people for the efforts done and not on the results achieved. Then only a team can remain cohesive.

Recognition should be done in time and not after a long time. People will be expecting someone to recognize them after fulfilling certain tasks or after completing certain process. The recognition need not be in monitory terms, but a word of appreciation, a word of support, or a word of consoling can do a big trick.

6.3.7   Morale

If the morale of the team members is high, everyone feels that he/she is a part of the team. They are enthusiastic in achieving the goal and shall be really happy when goals are achieved. They will not think anything against the interest of the team. When morale is good the members will not get disappointed with small failures as they strongly believe in their team and the efforts made and are confident of success in next attempt. A leader should always ensure that the morale of the members is not allowed to drop down.

6.4   Process of team building

There are four stages in team building (Fig. 6.2). When a team is formed, the members are new and the leader has no clarity on the competency of each member. Each member is enthusiastic as he is involved in the team and the morale of the team is high. This stage is called as orientation stage. During this stage, the members introduce themselves to other members and also share their previous experiences, achievements and assure their support to other members for fulfilling the task of the team. The leader allots different tasks to members considering their capabilities as described during orientation. When the work starts the problems come to surface. The members have to face various difficult situations depending on the nature of activity and start blaming other team members for the failures in accomplishing the task. Some dissatisfaction creeps in the minds of the members and the morale comes down. This stage is called as dissatisfaction Stage, and the role of leader becomes very important. Any team formed to accomplish a challenging task has to pass through this stage of dissatisfaction. If a team claim that there was no dissatisfaction, it normally means that the work was not started or the target taken was too small and not challenging. Leader needs to understand the problems of each member and workout a strategy to overcome that problem. He may have to reallocate the activities or combine certain activities or even eliminate them. He might have to take some new members depending on their skills and capabilities and remove some those cannot contribute for the task undertaken. This stage is called as resolution stage. Once the resolution is proper, the team starts working and gets the production. The working of the team shall be smooth and each member starts contributing as per his capacity. This stage is called as production Stage. The morale of members increases as the production starts increasing.

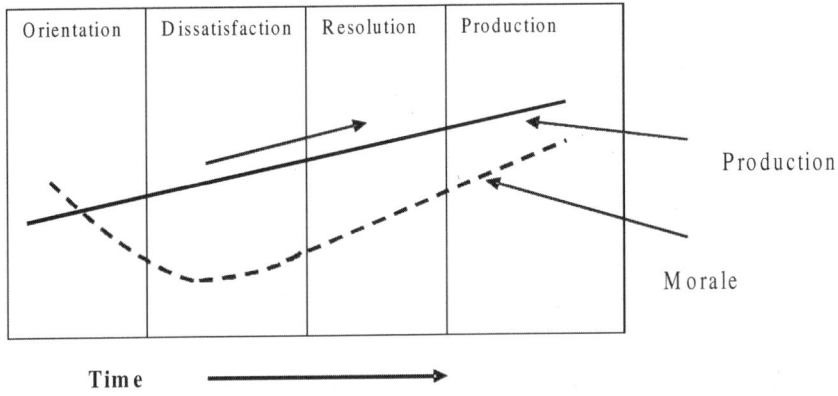

6.2 Stages of team building.

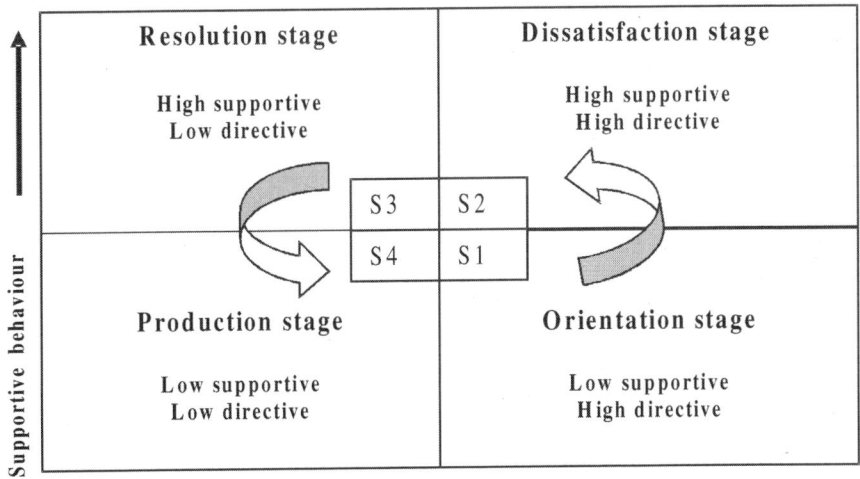

6.3 Leadership behaviours.

The team leader has to exercise different styles of leadership depending on the situation (Fig 6.3). He shall be low supportive and high directive in the orientation stage, whereas high supportive and high directive in the dissatisfaction stage. When the team reaches resolution stage, the leader shall be high supportive and low directive, whereas when the production stage is reached, he shall be low supportive and low directive. The team members will perform by themselves and there shall be no need of additional efforts by the leader. The leader shall have time to plan the further activities.

The success of a team depends on the leader. He should be in the team along with the members, holding the team members in their respective positions as a thread in a garland, which is not visible from outside, but holding all the flowers in their respective position to make the garland beautiful. The leader should be motivating the team members to get the results on a continual basis.

6.5   Motivating a team

Motivating is stimulating the interest of a person in an activity and getting him to do it. Psychological studies show that normally human beings make use of only 10–15% of their ability. When motivated, the talents and strengths hidden are exposed. Achievements develop confidence among persons. Motivation helps a person to grow.

Everyone has a greater potential than he is exhibiting, and as a leader, it's your responsibility to maximize the potential and performance and the results of each member. Creativeness often consists of turning up what is already there. To turn people on, you must first find out what they really want, and then, show them how to get it. To motivate others is one of the most important management tasks. It comprises the abilities to understand what drives people to communicate, involve, challenge, encourage, set an example, develop and coach, obtain feedback, and provide a just reward.

Each one has some ambition; if he feels that the work done shall help in reaching his goal he shall be motivated. Everyone has one or other problems; if he feels that the work he is doing can help in solving the problem he shall be motivated. Everyone has some wishes; if he feels that the work can help him to have his wishes, he shall be motivated.

To become a motivational leader, you start with motivating yourself. Motivate yourself with a big vision, and as you move progressively toward its realization, motivate and inspire others to work to fulfill that vision. If you are not motivated, you cannot motivate others. Inspirational leaders create an inspiring culture within their organization. They supply a shared vision and inspire people to achieve more than they may ever have dreamed possible. They are able to articulate a shared vision in a way that inspires others.

Motivating the team members is not a onetime task, but a continuous activity of a team leader. There are different methods of motivation like offering incentives, providing facilities, and so on, whereas, a leader should not assure anything that is not in his control. The following guide might help a team leader in motivating his team members.

Deliver an inspirational opening by relating an example of teamwork, using a quotation, telling a personal story or by reading a testimonial. The inspirational opening can help increasing the morale of the members and inspire them to be active members. However, once the work is started, the same speeches cannot help motivating the members. As the work goes on the leader should hear the reports of achievements and congratulate the winners. The areas of growth are to be noted and made public. The non members also should know about the achievements of the team. Lack of confidence demotivates the people. Filling confidence among the followers is one of the best motivating tools. People live because of the hopes. You should ensure that their hopes are not shattered.

When people feel that they are insecure, they lose interest in the job. Even if we offer a very high salary package, they will be spending their time more in searching for an alternative job where there is security. A good leader shall ensure that his people are secured. A leader needs to support his followers to get support from his followers.

As the team develops, the challenges are to be addressed and the team should identify the opportunities. Creative thinking sessions can help in identifying more opportunities which also motivates the members to be more active.

The goals set are to be SMART, i.e. Specific, Measurable, Achievable, Realistic, and Time phased. Non achievable or non realistic demoralize the members and the team becomes ineffective. Once the team goals are fixed, individual goals may be set. Identify the skills needed to achieve goals and make efforts to impart those skills among the team members. Once the skills are developed, the commitments may be taken from each member or team as a whole.

Whenever the team accomplishes the task, the leader should make an inspirational close so that the people are energized to take next task.

6.6   Interpersonal conflicts

The conflicts are one of the main reasons for the downfall of an organization, and hence, the supervisor has a responsibility to avoid conflicts as far as possible. It may be due to a difference of opinion relating to a particular matter. It may be a fight or struggle or clash of opposed principles or opposition of incompatible wishes or needs in a person.

The causes for conflict are many like self interest, misunderstanding and lack of trust, different assessment, and low tolerances for change or implementation. Self interest is one of the major causes of conflict, which may be due to any of the following:

- Feel of insecurity by implementing certain system or technology.
- Feel of insecurity by promoting or giving importance to certain individuals.
- One might be aspiring for certain benefits or position, but deprived of it because of certain decision taken by others.

Misunderstanding and lack of trust are the outcome of previous experience either with the same person or with similar situation. People do not trust the man proposing or initiating a change, feeling that figures projected are doubtful and losses are more than gains.

Different assessment can be made for the same situation by different people because of their perceptions and experience. When the assessments are different and big differences found when the situation was assessed by others it leads to a conflict.

When the tolerance for change/implementation is low, people feel that they will not be able to implement. In such cases, they oppose the very idea of the project which leads to a conflict.

6.6.1   Areas of conflicts

The conflict may arise in any area of an activity like technology adaptation, fixing of targets, allocation of people for different jobs, product and market segmentation, allocation of funds for different activities, sharing of profits, bonus and dividends, sharing of powers, assessing performance and recognizing the performers, affiliation/supporting political/trade unions and so on. The conflicts become interpersonal when it is taken on personal issues rather than on company matters. Interpersonal conflicts are mainly due to

- Improper communication leading to confusion.
- Taking decision without taking other person into confidence.
- Knowingly or unknowingly using a word or referring an instance that is not liked by others.
- Influence of a previous experience where one person got more benefit or made the other to loose by his tactics.
- A feeling that other man as responsible for the personal loss of a person.

Steps recommended to reduce interpersonal conflicts are transparency in the systems and procedures, respecting and adhering to the established systems and procedures, involvement of users and the stakeholders in decision making, periodic and impartial reviews of all the activities and listening to the grievances of the people honestly.

6.6.2   Dealing with resistance

There are number of methods of reducing the resistance to implementation of any project or proposal. Participation and involvement of the people concerned is the first step to be taken to deal with resistance. When people are really involved, there shall be no resistance for the basic purpose, but there may be resistance for certain acts. By facilitating and providing support, the resistance can be further brought down. Sometimes, we may have to negotiate and make people agree for the proposed project by offering certain benefits. In such cases, normally written agreements are made so that the problem should not be posed again and again. Sometimes, manipulation of certain figures or steps may be needed to get the cooperation of the people.

Explicit and implicit coercion is the last resort in dealing with a resistance which may be by referring the matter to higher management or HRD and creating witnesses of senior/respectable persons. Most important aspect in dealing with resistance is to establish yourself as a friend and well wisher by your deeds. When people feel that you are not an enemy, but a friend, they shall join hands with you in making your project a success. Help the people when they are in trouble; they shall cooperate with you when you need them. Be impartial all the time. Avoid imposing your ideas first, but make people understand and like your ideas. Following are some tips to reduce inter personal conflicts:

- Give respect and recognize the strengths and achievements.
- Do not under estimate others.
- Have self control and be patient.
- Do not attack/blame a person but attack a system.
- Do not criticize but explain and convince.

Whatever might be the issue, whoever may be involved, whoever might be wrong, the loser is the organization in every interpersonal conflict. Therefore a supervisor must take all precautions to avoid the interpersonal conflicts in his work area.

6.7   Quality circles and project teams

Quality circles are team of people from a section working in a similar environment doing identical jobs coming voluntarily together to solve some of their work related problems. The concepts of quality circles were developed and started at Japan in 1950, by Dr. Ishikawa. The very important aspect is creating an environment by the top management that workers feel the importance of voluntarily coming forward to solve their work related problems. The faith in management by the workmen is the driving force, and not any financial incentives.

The transparency in the top management actions and policies shall influence the people to come voluntarily forward to take some of the load on themselves and work for solving the problem. We cannot form quality circles by force.

The main idea of a quality circle is to make people work as a team, and not solving the problem.

Normally, a quality circle consists of 5 to 9 people, and they meet periodically and discus the problems and initiate corrective actions. Once they get the results, the same is presented to the top management, who recognizes the efforts and rewards the teams suitably so as to increase their morale.

The quality circles are self-evolved voluntary teams who identify the problems themselves and try to solve them. This cannot solve bulk of the problems that are either technical related or system related, which needs specialist teams to work for that specific task. Normally, 80% of the problems need management interference and 20% can be handled by quality circles. If a problem is found as important, and the quality circles have not identified that as a problem, the management shall have to find a way to solve them.

Any problem shall have its roots spread at different places, and hence to solve such problems, we need to take the help of concerned people. Hence, a cross functional team is needed. The management nominates these teams. The team members shall be assigned specific authorities and responsibilities.

The cross functional teams or project teams shall have a time limit to complete their jobs, and after that time is over, whatever might be the position, the team shall be dissolved, and a fresh team is formed to attack the remaining part of the problem.

The teams follow various steps of problem solving, viz. problem identification, observation, root cause analysis, devising alternate actions, taking suitable action, verifying the results and reinforcing the systems.

The team members in a project team shall be experts in their area and by a combination of knowledge and experiences, along with their authoritative positions, the problems shall be solved.

7
Decision-making process

7.1   Importance of decision

Taking decision is one of the very important tasks of a manager or a supervisor. Decision and planning go together. Without a decision work cannot start. Without starting work cannot be done. Without a decision work cannot continue and also cannot complete. A decision gives green signal for the people to move forward in a direction. It gives directions for the people to move.

The decision may be appropriate or may not be appropriate for the situation, but one has to take a decision. World can tolerate wrong decision makers but not non-decision makers. You need to decide your future, your activities and your goals. The success of a man depends on the timely decisions taken by him.

7.2   What are to be decided?

In an industry, we need to decide on number of things starting from mission, vision, objectives and goals, capacity to be installed, machines to be procured, men to be employed, organization structure, authorities and responsibilities, work methods, controls and checks, actions on non-conformities, actions on problems, other investments, products to be manufactured, process to be adopted, machines to be engaged, maintenance activities to be done, and so on. There is no area where decision is not needed. When we consider the supervisory staff, they need to take decisions on engaging available workers on different activities depending on the skills, balancing the production activities by looking to stocks, employee availability, machine availability and customer requirements, taking corrective actions for deviations in quality and productivity, replacement of parts, prioritizing the activities as per customer needs, and so on.

7.3   Types of decision

The decisions may be grouped into short term, medium term and long term. The supervisors are normally supposed to take short term decisions to get the results in each shift, whereas the top management is supposed to take long term decisions in line with the company policies, market requirements, economic situations, etc. Following are some of examples:

Short term

- Allocation of machines for different counts, colour combinations, sorts, styles, and brands working considering the lot completion positions.
- Allocation of colour codification depending on the number of counts running and the availability of colours.
- Purchase of materials that were short or needed for emergency.
- Allocation of work basing on the people and machinery and the skills.
- Stopping the machinery considering the stock position and giving it for maintenance activities.
- Assigning petty works to the people in case of stoppage of machines or process due to any reason like power failures, fire accidents, short of materials, excess stock and delay in decision by production planning etc.
- Corrective actions to be taken for deviations in quality.

Medium term

- Deciding annual and quarterly product mix and production planning
- Procuring of materials for orders received.
- Planning for spare replacements and related maintenance operations.
- Deciding on minimum stocks of spares, consumables and other materials.
- Contracting for our house activities for particular order.
- Bonus and dividend distribution

Long term

- Capital investments
- Policies and objectives
- Wages and promotions
- Deciding broad product mix on which the capital investment is based.

7.4   Factors influencing decisions

There are number of factors influencing the decisions; for example, market trends, availability of products in market, culture and fashion, expected changes in the demands, purchasing capacity of people, understanding capacity or knowledge level of people, changes in government policies, financial position of the organization, availability of funds, availability of human resources, development in technology, pushing and pulling forces within organization, long term and short term goals, performance of competitors, expansion and modernization plans of competitors and so on.

When we consider the decisions to be taken by a supervisor, the main factors that influence will be the stock of material in process at various level, the materials produced and to be produced in each variety as per the purchase order of the customer, the availability of men and machinery, the temperature and humidity levels and its effect on the working, the availability of cans and bobbins of different colours, the packing materials, the storage facility available, the working problems and the quality levels, emergency issues like fires, machinery breakdowns, accidents, product changeovers required, and so on. There is no single formula to guide the supervisors for decision making excepting *"See the situation, have presence of mind, think and decide"*.

7.5   Base for decisions

Although there are various factors influencing the decision, we need to have some base developed for taking decisions. They are facts and figures, current situation, trends, financial implications of different solutions, expected resistance, people those are resisting or influencing and probability of success or failure.

Factual approach for decision making is very important, as it prevents from taking wrong decisions in majority of the cases. Understand the current situations, study the trends and decide on your strategy to achieve the results. In the process keep the interests of the stakeholders in front and work for fulfilling them. The stakeholders for any business organization are the customers, employees, shareholders, top management, suppliers, government and community.

Success is the result of number of attributes, whereas timely decision is the most important one (Fig. 7.1). The other attributes are workmanship, knowledge, determination, creativity, physical ability, values, courage, dedication at work, thinking ability, obedience, analysing capacity and so on.

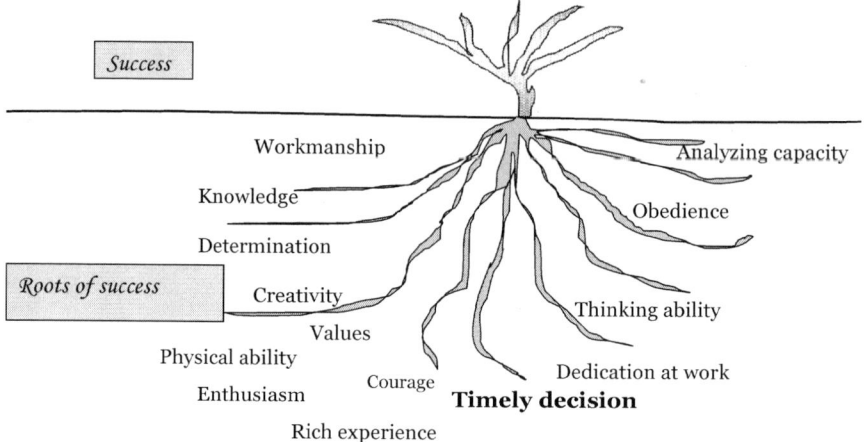

7.1 Roots of success.

7. 6   Tools for decision making

There are numbers of tools that can be used to take suitable decision; may be statistics based or logic based or a combination of both. The tools mentioned below do not give any inference about the decision to be taken but help in understanding the situation. They are grouped either in product and process design tools, management tools, quality control tools and operation research techniques. Hazard Analysis and Critical Control Points (HACCP) helps in identifying the possible hazards and the actions required to be taken. The Failure Mode Effect Analysis (FMEA) explains as to which failure can prove very costly or dangerous so that actions can be taken on priority to prevent that failure. Quality Function Deployment (QFD) helps in identifying different parameters in process those can be combined together to get the required results to suit the customer needs. Process Design Programme Chart (PDPC) tries to identify possible hurdles in the process so that suitable preventive measures can be taken. The relation diagram helps in identifying complex relations between two activities whereas; an affinity diagram indicates the natural relation between the available data. The tree diagram helps in mapping an activity and identifying goals and sub goals at each level. An arrow diagram, which is considered in both management tools and in operations research, helps in critically analysing the process and various events and their interactions. This helps in scheduling the activities. Boundary analysis, which is considered as one of the QC tools, helps in understanding the limits of each activity so that designing the activities and analysing the failures can be made.

Seven QC tools were identified by the team of Dr. Juran, Dr. Deming and Prof. Ishikawa in their total quality movement at Japan in 1950s; whereas number of tools were identified and added later. Now there are various QC tools which can be used for problem analysis as well as decision making. The popular QC tools are data collection, brain storming, flow charts and process mapping, critical activity chart, boundary analysis, check sheets, concentration diagram, stratification, run charts, stratified run chart, cause and effect diagram, pareto analysis, histograms, scatter diagrams, force field analysis, spectrogram and so on.

Apart from QC tools, the operation research techniques like linear programming, game theory, programme evaluation and review technique along with critical path analysis, sequence analysis, queue theory, simulations and breakeven analysis also can be used as tools for decision making.

7.7   Decision-making journey

Various tools and the data can be used for taking a decision, but one need to follow certain path. The journey for taking decision is through various steps starting with the diagnosing of the situation before taking decision.

7.7.1   Diagnosing the situation before taking decision

Diagnosing the situation before taking decision is the first step in the journey for decision making. It involves analysing the symptoms and formulating theories linking the symptoms to the effects seen. The theories so formulated are to be tested by various means like use of current operational data, cutting new windows when needed to facilitate verifying the theories, designing and conducting the experiments and finally identifying the root cause for each happening.

7.7.2   The journey

The journey for decision making has number of steps as follows:

- Identify the area where decision is needed. When and where to take the decision is very important before deciding on what decision to be taken.
- Observe the situation and analyze the factors influencing the present situation.
- Find the key contributors and their share for achieving the objectives from the present situation.
- Develop alternative solutions for achieving the results.

- Choose from alternatives that can be effective, economical and easy to implement.
- Anticipate resistance; any action or decision shall have one or the other resistance. It is needed to anticipate resistance and decide preventive measures.
- Plan alternate steps; there might be other methods of achieving the goals. One need not stick to one procedure or one process only to achieve the result.
- Implement the decision – The result is the result of an action taken as per plan and not of the action planned.
- Establish control – Any process needs to be controlled to get the sustainable result.
- Review the effectiveness of the actions taken.
- Amend the decision if needed. Our goal is to achieve the goal, and the path adopted is immaterial as far as it not unethical or illegal.

7.7.3   Distracters for decisions

While taking a decision, one should be careful about certain distracters those lead to wrong decisions. Normal distracters for good decision making are as follows.

- Dumping of unconnected information. This is a major problem in recent years because of the huge capacity of data collection and storage by using computers. Out of the data stored, about 5% or 10% might be useful for taking the decision whereas the other data shall distract decision. One should be clear in filtering the data and verifying only the required data for effective decision making.
- Influence by people in power is a major problem. The people in power want the decisions to favour their personal interest rather than the organization's interest or the country's interest. A technician becomes helpless in such management as he cannot survive by going against their interests. When the decision fails, the technician is made the scapegoat.
- Previous experiences leading to bias is one of the distracters of good decision. The previous experiences always try to influence the decision maker but the present situation might be entirely different than the previous one.
- Success or failures at other places distract the thinking of the decision maker. The decision taken at other place, if successful, shall be taken as a decision here also although the situations are different.
- Offers and bidding by competitors forces the decision maker to compromise at second or third option rather than the first option as per his decision.

- Unstable government and government policies forces the decision maker to take certain decisions although he is not in favour of that.

7.8   Work process designing

The work process designing is very important among decisions to be taken for getting the required quality and productivity. It involves the installations and aligning of various machines and mechanisms, deciding on their settings and other process parameters, deciding the control atmosphere to be provided at work place, establishing measures to control the process, the balancing of the process, determining the required competency of the people to work and arranging for proper education and training, planning the required support activities like maintenance, inspection and testing, and so on.

Normal questions asked while deciding on a process are as follows:

- What is expected out of that process?
- What are the steps involved?
- What are the interactions with other processes?
- What are the factors contributing for the success of the process?
- What are to be checked and controlled?
- Which are the value adding points and what are cost adding elements?
- What are avoidable and what are essential?
- How much to be controlled?
- What data and information are required for control?
- Who should collect data and analyse?
- How the effectiveness is to be measured?
- How it is to be monitored?
- Who should monitor and review?
- What should be the frequency of checks and reviews? And so on.

A flow chart may be made showing the process flow (Fig. 7.2), the controls and checks needed at each process, the interactions and decisions to be taken at each step so that the decision is explained clearly to the implementers.

7.9   Loyal friends in decision making

Decision making can be effective when we understand all aspects relating to taking decision. The following questions are to be asked again and again to have clarity and for taking effective decision. The questions are:

- Why? Why this work is needed to be done?

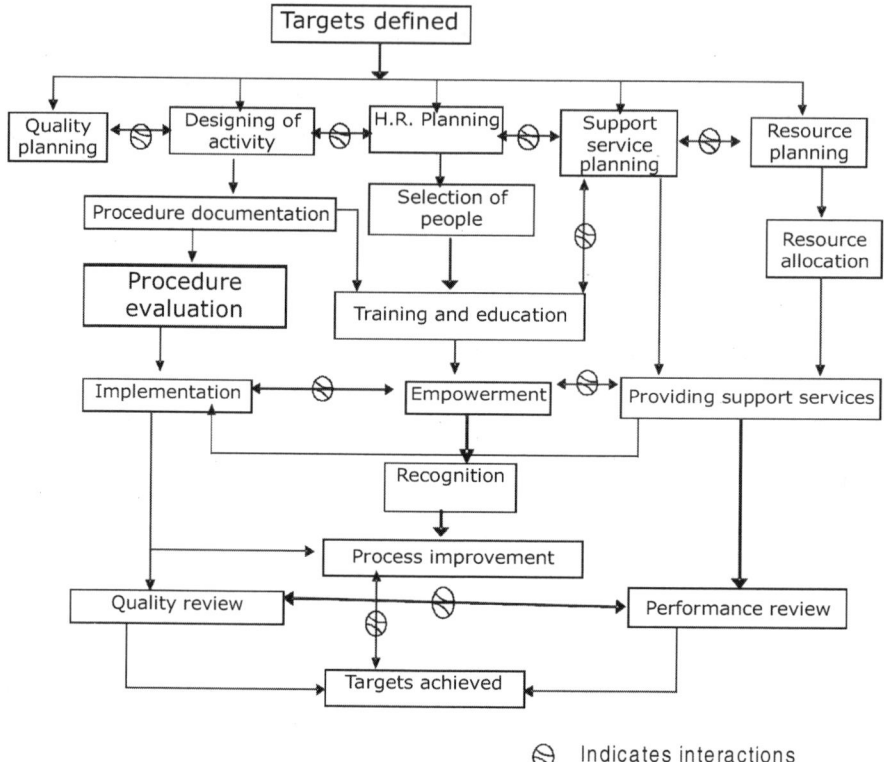

⊖   Indicates interactions

7.2 Work Process designing.

- What happens if this work is not done?
- Where? Why there only and not at other place?
- What? What is to be done? What is to be achieved? What is the purpose?
- Who? Who should do? Why not others?
- When? When to do the work? Why at that time only?
- How? How the work is to be done? Why not in different way?
- How much? Why not more or less?
- So what? If it is not done what happens? What is its effect? How it is going to affect our decision or the working?

The six loyal friends "why, where, what, when, who and how" should always be with us helping in taking correct decisions.

Communication and supervisor

8.1   What is communication?

We have number of emotions and messages to be conveyed to others so that they can understand and work accordingly or cooperate with us in doing our activities. Some times the message to be conveyed may be in the interest of the receiver and giving the clarity in the message is very important. The messages may be of personal nature or of the organization or the society in which we are working or living. Conveying the message effectively is termed as communication.

Communication can be explained in different ways. It is the art of conveying the message or information. Communication is a social intercourse and no living beings can survive without communicating their feelings or thinking to others. It is transmitting message effectively by any means like speaking, writing, non verbal expressions or by doing. If the receiving person has not understood the message clearly, it is not a communication.

A supervisor is always dealing with human beings and his success depends on how he communicates with his people and gets the works done. He has to communicate number of messages to his fellow workers or team members like the company's policies, group targets, the procedures of performing a job, works to be done for the day, the points to be checked while working, the complaints and feedbacks received from customers relating to their area of work, the quality levels of the materials produced by them and actions needed to improve the situation, the production achieved and the shortfalls, and so on.

Employee communication provides a platform to employees to interact with his/her administrators for various information as well as support or update requirements. It helps in ensuring that data updated is correct, there is correct understanding of processes, policies and practices, information is quick and authenticated.

8.2   Methods of communication

Messages can be communicated in different ways depending on the situation. It may be

- Verbal communication
- Written communication
- Advertisements and mass communications
- Nonverbal communications

8.2.1   Verbal communication

Verbal communication may be one to one or in a small team is one of the most effective methods of communication provided the person listening is attentive and the one explaining explains clearly. However unless it is verified one cannot ensure that the communication was proper. The person communicating shall be of the opinion that he has communicated the message fully, whereas there are chances that people have not listened and understood the message. The verbal communication methods include talking personally, talking on phone, giving a lecture to a small group, addressing masses, discussing with in a group and making public announcements

Talking personally is more effective compared to all other means of verbal communication. One can make communication better with suitable body language while communicating in person, whereas the same is not possible while conversing on phone.

Talking personally or on phone are two way communication whereas giving lecture, addressing masses and making public announcement are one way communication, and there is no clarity as to who got the message and who did not. In two way communication one can ensure that the communication was effective by cross checking with different questions. Although discussing in a group is a two way communication, we are not clear as to whether all are actively participating or only a few are active.

8.2.2   Written communication

Although verbal communication is very effective, there are chances of people forgetting what they talked depending on the other works or priorities the mind has. The interpretation may not be correct in one hearing and one might prefer to refer second or third time. A written communication shall ensure that the contents are not forgotten as the message can be referred any number of times and interpreted with different angles.

Written communication is more useful while communication is formal, and needs an evidence of conveying the message in time to the concerned. There are number of means for written communication. They are writing personally, sending e-mails either to a single person on personal ID or by spam messages, displaying on notice board, amending the documents and notifying and displaying slogans in work area

Writing a personal letter has more weight compared to all other written communications. It gives a personal touch and with some emotional connections. People respect personal letters written to them rather than a general note displayed on a notice board or an announcement in papers.

When the communication is to be made to masses about a change the notice boards and news papers are used. Although spam messages in e-mail also has the same intention, readers normally do not respect the spam message because the tracing the sender is difficult as he operates from a virtual location.

Notice boards are meant for conveying messages to a particular group of people working or staying in a known limited area. Public notice given in a news paper does not have the same effect as that of a notice put in the office notice board. Displaying slogans in work area is only for the purpose of giving awareness, whereas the notice put on a notice board is for action.

A supervisor in a textile and apparel industry has to communicate in writing the following in order to avoid confusions and to have record for reference.

- Giving a slip with changes to be made in case of changes in count, sort, lot or style to the concerned person implementing the change.
- Giving instructions for preparing recipe for sizing, dyeing, printing or any chemical processing activity.
- Giving instructions to maintenance person for replacing parts or change in settings.
- Giving instructions to the next shift supervisor by writing the same in log book.
- Giving requisition for the materials to be brought from stores.
- Giving material return slip to the stores while returning excess materials back to the stores after completion of a style or lot.
- Advising the human resource section in case of any misdeed done by any of the fellow employee.
- Leave application and acknowledge with the sanction or refusal of leave.
- Permission for an employee to go out of the office.
- Employee grievance if any.
- Action taken reports to be submitted to the top management for specific actions suggested by top management.

8.2.3 Advertisements and mass communication

Advertisements and mass communication are one way communication and hence there is no clarity as to who got the message and who did not. Therefore, these are to be done repeatedly from time to time. There are different types of advertisements and mass communication systems as advertisements in news papers, street banners and leaflets, propagating in seminars, conferences, gatherings, writing books, articles, novels, poems and dramas and audio visual presentations in television and movies, street plays and dramas and radio.

Advertisements given in a news paper can reach to large masses depending on the circulation of that paper; whereas street banners and leaflets are limited to certain areas. However, street banner can attract more people than an advertisement given in a paper because of its size, art work and colour combinations. The leaflets are normally not kept by the people as they throw it out, whereas a street banner shall be seen number of times when people are moving in that street. If people are interested, they may keep the news paper advertisement with them or else it shall be forgotten.

Propagating in seminars, conferences and gatherings are done while communicating message to like minded people. Writing books, articles, dramas, poems, novels etc., can communicate message to the interested people, but we cannot say that all the readers got the same message. The understanding of the message depends on the mindset of the reader, his perceptions and experiences, the culture and so on.

The audio visual systems of mass communication are more effective as they can attract and keep the viewers together and message can be conveyed with illustrations. In a textile mill or a garment factory audio visual presentations are recommended for educating the people on good work practices, methods of assembling and dismantling, safe methods of material handling, awareness on safety aspects at work, reasons for defects and their remedies, awareness on health and hygiene and awareness on the welfare measures taken by the company.

8.2.4   Nonverbal communications

The verbal communications and written communications do not convey the hidden meanings or intentions of the people who is making the communication. A person may be praising a person to get favour from him by sweet words or by writing a sweet letter, but by observing the nonverbal communications that person is giving, one can try to understand the real intentions. There are number of methods of nonverbal communication.

Following are some forms of nonverbal communication.

- Body language – facial expressions, expressions by eyes and body movement
- Voice modulation – firm, irony, casual or joking
- Being a role model
- Sign boards / cartoons

By observing the person communicating, his eyes and body movements, facial expressions, it is possible to know whether the person is frank or not. It requires systematic practice to understand the meanings of such body postures and movements.

In textile mills, because of the noise and also the distance, people cannot come and talk personally all the time but shall be sending their messages by signs. These signs have specific meaning that is understood only by the selected people for whom the message is being sent. The supervisors should get themselves trained in understanding the nonverbal communications they are getting so that they can understand the reality and take actions.

Some examples of non-verbal communications indicating that the people are not interested in your words are the people yawning when you are explaining, people seeing other side or attending to other works, receiving or making phone calls, going out of the room with some reason, seeing watch again and again and coming late when called. When people are interested in you, you can make it out by their eyes, which shall be seeing your eyes without any break.

8.3   Effective communication

Communication is said to be effective when the other person understands our message in total and acts as per message. We need to verify whether our communication has reached the people or not by various means.

- Do not be happy just because you have conveyed, but confirm with the other person that he has understood it. You can ask the person to explain back the same to you as you explained to him, or ask that person to explain the same to another person in your presence. In a number of times we see that people can explain only 50 to 70% of what we explained and not 100%. This percent goes on reducing as the same message is conveyed from person to person in a line.

- In number of cases, there might be a difference in meaning for the same words at different places. Therefore, the message given by you might be understood in a different way. For example, a general manager of a spinning mill suggested the purchase section for increasing in length of cotton, which promptly purchased long staple cottons. The meaning of longer length as per general manager was to have more stock of cotton so that same mixing can run for a longer time.

- You are an expert in the subject you are talking, but the one who is listening is not. You may feel that you have conveyed everything, but the receiver might not have understood even 10% of what you said.

● It is seen that human mind works faster than ears or eyes, and people start thinking further when we are explaining something. They are not listening, but are preparing some plans or dreaming something by linking to our previous words. We cannot say that they are not interested, but they are over interested and more enthusiastic. Because of over enthusiasm they are not able to listen fully.

Communication is said to be effective when the required information is delivered to the concerned in time and the receiver understood the message clearly and started acting as per the message. The supervisors normally communicate with workers who are less educated, having lesser exposure to various fields. In number of cases, the mother tongues are different between the supervisor and the workmen, and hence proper expressions might be missing. Therefore, there are always chances of not fully understanding the communications given by the supervisor. If the supervisor has to become successful, he has to ensure that his communications are effective. One should remember that

⇒   Successful bosses have good communication skills.
⇒   They learn from people, including their employees.

Costing and cost of quality

9.1 Definitions

Costing and cost control is an essential activity in which supervisors have to take an active role. Before discussing on the methods of costing and cost of poor quality, let us try to understand some of the terms normally used.

- *Cost* – The monetary value of resources used or sacrificed or liabilities incurred to achieve an objective, such as to acquire or produce a good or to perform an activity or service.
- *Costing* – The system of understanding the costs incurred for performing an act or to produce a product or service.
- *Elements of cost* – The costs incurred for various materials and actions are grouped as elements as costs; they are material costs, employee expenses (labour cost), power, fuel, interest, depreciation, stores, overheads and administrative expenses.
- *Cost of quality* – All the cost incurred to produce and deliver the goods and services of correct quality right at the first time. This does not include the normal manufacturing costs like material cost, stores cost, labour cost, power cost, fuel cost, overheads etc.
- *Cost of poor quality (COPQ)* – Cost of poor quality consists of those costs which are generated as a result of producing defective material. It is the amount of money a business loses because its product or service was not done right in the first place.

9.2 Elements of cost

Controlling the costs is very necessary to be competitive in the business. The management is to be provided with the data for cost control with analysis and classification of costs in different elements. The classifications has to be made to arrive at the detailed costs of departments, processes, production orders, jobs or other cost units. One can find the total cost without classifying, but it will not help the management to control the costs and reduce unwanted or non-productive expenses. Although the cost accountants have the responsibility of computing the data for costing and submitting to management, the technical staff should have the basic knowledge so that they can take appropriate actions in reducing or controlling the costs.

When we talk of a textile mill, we can group the activities as production, engineering, purchases, stores, marketing, administration, human resource management, quality assurance, etc.

The production activities can be sub-grouped as production planning and control, designing and sample development, maintenance of production machinery, maintenance of sub-stores in production area, material handling activities, production activities and packing activities.

The Engineering activities can be sub-grouped further as mechanical, electrical, civil, electronics including communication and information technology and so on. There can be sub-grouping of the activities as follows:

- Mechanical – Work shop maintenance and repair activities, boiler house operations, vehicle maintenance, hydrant maintenance, humidification plant maintenance, cranes and hoist maintenance.
- Electrical – Power generation and distribution, maintenance of electrical equipments like motors, control panels, cables, lighting, air conditioner, etc.
- Civil – building maintenance, road and garden maintenance, water supply, effluent treatment, sanitation, etc.
- Electronics – computer and PLC maintenance, telephone, fax and photo copier maintenance, etc.

The purchase activities can be sub-grouped further as procurement of raw materials, product accessories, machine spares, maintenance accessories, colours and chemicals, fuels, packing materials and general items.

The activities in the stores can be grouped further as raw material godown, accessories stores, general stores, lubrication stores, diesel stores covering diesel, petrol and other inflammables, fuel stores like coal, fire wood and husk, chemical stores and stationary stores.

The activities of marketing can be grouped further as market data collection and market research, whole sale and retail sales, exports and domestic sales, market development, customer services, customer grievance handling and recoveries.

The activities at human resource management can be grouped further as recruitment, promotions, retrenchment and retirements, training, developments and transfers, welfare, compliance, security, time office, attendance and distribution of work force, leave management, discipline management, industrial relations and grievance handling.

The activities at quality assurance can be sub grouped as incoming material inspection, on-line production inspection, finished materials inspection, laboratory management and testing, process control studies, process setting up and work norm fixation.

The administration activities can be further grouped as finance, accounts, costing, insurance, wages administration, taxes, and liaising with government for license.

In each sub group, we need to analyze the expenses against each element of cost and work out the contribution of each element for the total cost of the product or service.

9.3   Methods of costing

There are different methods of costing. The systems normally followed in textile and apparel industries are historical costing, standard costing, marginal costing and activity based costing.

9.3.1   Historical costing

In this system of costing the past performance (three months, six months or even one year) is considered and cost is worked out. This system is suitable where single product is being produced. If the product combinations are getting changed, it shall be difficult to arrive at the cost per unit production. This system gives past performance cost correctly and can be used to study the trends in cost escalations. But it has the following disadvantages:

i.   It does not give cost forecast.
ii. It is a post mortem.
iii. Does not help management to take decision as cost is not differentiated as fixed and variable.
iv. If the products are getting changed frequently, it shall be more confusing.

9.3.2   Predetermined costing

In this system the performance of man, machine and material for basic period of three months, six months or one year is pre-determined and the predetermined cost is worked out. The trends in the historical costs are taken and the same trend is projected further and costs are predetermined. The advantages of the system are that it gives a cost forecast and can convince as trends of previous period are taken as a base. The disadvantages are that as the expenses are not split as fixed and variable, this system does not assist management in taking some decisions correctly. This system does not give cost control

9.3.3　Standard costing

In this system of costing the performance standards of man, material and machine are fixed up scientifically and the standard cost is worked out. The prevailing prices are considered while working out the costs. No previous trends are taken and provision is made for escalations. This system gives cost forecast and that can be used for excellent cost control. It can be worked for different product mixes. This can be used for preparing budget for the performance; however, it is assumed that the same condition of market and technology prevails. The main disadvantages are:

i.　This is expensive system.
ii.　As the expenses are not split as fixed and variable it does assist in taking certain decisions.
iii.　As the cost is not shown as fixed and variable, the variances are magnified.

9.3.4　Marginal costing

In this system of costing the variable expenses are apportioned to the product while fixed expenses are divided by total cost units of the main centre and expressed as minimum contribution to fixed expenses required per spindle per shift or per loom shift etc. It is very simple system of costing and gives a cost forecast. It assists the management in decision making and is more advantageous where product mix is changing very often.

This system has following disadvantages:

i.　It needs periodic revision; however, a periodic revision gives accuracy.
ii.　It does not give cost control as given by standard costing.
iii.　The accuracy depends on what is considered as fixed and variable expenses.

9.3.5　Marginal-cum-standard costing

In this system the fixed and variable expenses are split up and allocated to cost centres but unlike marginal costing cost control reports of standard costing are prepared to get the benefits of cost control. This system is a blend of marginal costing and standard costing.

The advantages of this system are that it is not as expensive as standard costing, gives cost forecast, and assists management in taking decisions and cost control. The main disadvantages are that it needs periodic revision. However, it cannot be considered as a disadvantages as the market prices and other expenses are so much fluctuating, and hence, this is more suited.

9.3.6   Activity-based costing

Activity Based Costing (ABC) is a method for developing cost estimates in which the project is subdivided into discrete, quantifiable activities or a work unit. ABC systems calculate the costs of individual activities and assign costs to cost objects such as products and services on the basis of the activities undertaken to produce each product or services. Identifying activities is a logical first step in designing an activity based costing.

Each employee's time to the different activities performed inside a company are noted down either by a survey or by a study by Industrial Engineer. The accountant then can determine the total cost spent on each activity by summing up the percentage of each worker's salary spent on that activity.

Advantages of activity based costing are as follows:

i.   By associating cost to the activity, a clear relationship can be established between sources of activity demand and the related costs.
ii. This association can benefit the distributor in determining where costs are being incurred, what is initiating the costs and where to apply efforts to curb inflationary costs.
iii. This can be of particular value in tracking new products or customers.
iv. Cost reduction by identifying the real cost and taking action.

The limitations of this system are as follows:

i.   Data collection and appropriately distributing the overheads is very difficult.
ii. The data studied might not be real in all the cases leading to errors.

9.4   Cost of quality

The cost of quality is the costs incurred for producing the right quality right at first time and every time. The costs explained in the cost elements are required for producing the material, either it is of correct quality or poor quality. We need to understand the expenses made specifically to produce good quality. Let us take some examples.

What is required to produce good quality garment? The answer is quality materials [fabrics, accessories], trained workmen and well maintained machines, clarity in techpack and instructions, good working conditions, sufficient time and adherence to systems. The managements claim that they have provided everything demanded for the production, and still is not getting the quality products. What extra is needed?

Quality does not mean some thing extraordinary like gold plating, golden image or gold. Quality means conformance to specifications or fitness for the purpose for which it is made. If the product or services are not conforming to the requirement of user, then it is a poor quality. In other words, the product or services that could not satisfy a customer is a poor quality. When we talk of garment manufacturing, the quality if achieved should result in the revenues to the company. If we are not getting the returns for the activities done, it is a waste, or from the customer's point of view, a poor quality.

The studies of experts indicate that manufacturing operations commonly have a cost of quality equal to one quarter of their total turnover. It might vary from 15% to 35% depending on the culture and the systems employed. If the required quality is not achieved, then the investment shall become cost of poor quality. Therefore, the "cost of quality" becomes "cost of poor quality" if the things done are wrong. The cost of poor quality include cost of screwing up or repairing, the cost of reprinting a job because it was printed out of register, the cost of counting incoming material, the cost of inspecting any part of the job at any stage of production, the cost of late delivery, the cost of handling customer complaints, the cost incurred for conducting quality control programme etc., It may be noted that bulk of quality costs are associated solely with defective products or services, the cost of finding, repairing, or avoiding defects.

A realistic approach to making the supervisor aware of the necessity for quality is to outline the cost of poor or inadequate quality. Such an approach may be based on:

a) Basic cost of re-work which include cost of returning work, cost of operative time, cost of supervisors' time and cost of re-inspection
b) Loss of production because of operatives repairing defective work.
c) Loss of production by a likely quality problem.
d) Late delivery because of rectification time.
e) Customer dissatisfaction with quality.
f) Cancellation of orders leading to imbalance in production programme.
g) Reduction of profit because of loss due to excessive rejects.
h) Any other points which may serve to emphasize the picture.

9.4.1   Visible and invisible parts of COPQ

The COPQ can be compared to an iceberg, in which around 10% are visible and 90% are invisible in the water as shown in Fig 9.1. Among the costs of poor quality the reworks done, the rejections made and the complaints received are easily seen and hence called as visible part of COPQ. The other parts, which are very crucial but cannot be seen easily are the missed schedules, warranty claims, over staffing, over specification, over design, under utilization of workforce, stock holding, production capacity tied up, lost business, bad debts, lost image etc.

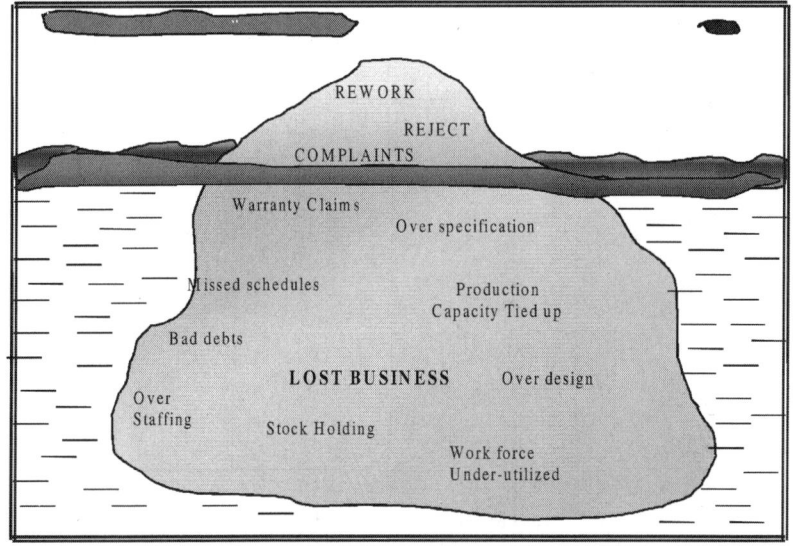

9.1 The visible and invisible part of COPQ

9.4.2   Good, bad and ugly costs

The components of quality costs are grouped as preventive costs, appraisal costs and field failure costs, which are often referred as good, bad and ugly costs. Figure 9.2 shows the components of COPQ. First there are costs associated with attaining or setting an adequate quality standard. They are incurred largely in advance of production. Insufficient money spent at this stage, for example, design and development may well give rise to unnecessarily high costs later. The second group is costs associated with maintaining an adequate quality standard. These are the costs associated with keeping the manufacturing and buying functions up to the quality specified in the design. The third category costs are associated

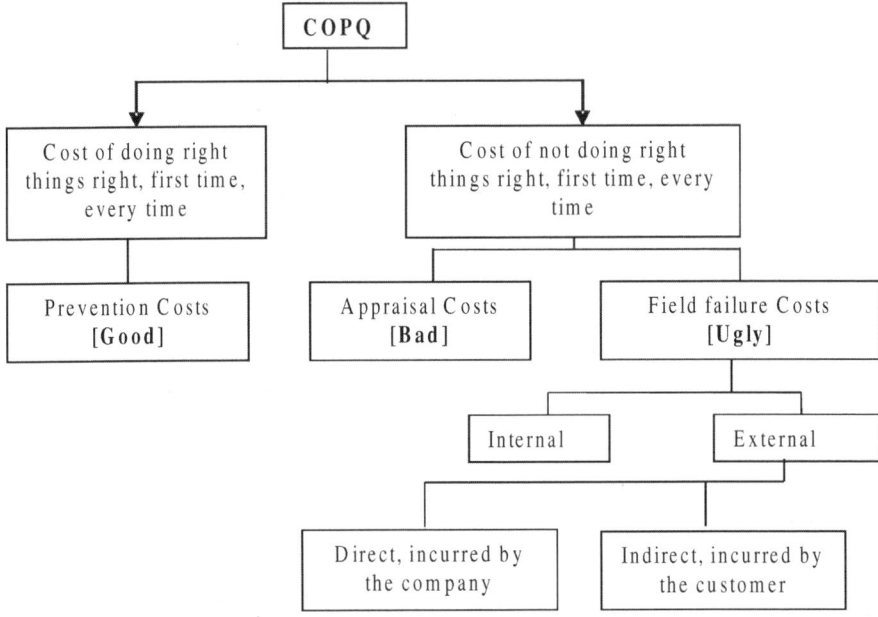

9.2 Components of cost of poor quality

with putting right any departure from standard. These include the costs of scrap, reprocessing, and claims. They are the costs, which arise as a result of shortcomings in, or insufficient expenditure on, the other two phases. They may be caused either by poor design, poor product engineering, and poor operative training or, by bad workmanship, or slips in inspection at the appraisal stage.

Preventive costs are good costs since such expenditures are meant to prevent errors from being made or, help employees do their jobs right every time. Hence, prevention costs are not really costs; they are more in the nature of an investment in the future. Typical prevention costs are incurred in activities such as preventing a problem from occurring and recurring, implementing the improvement process, quality related training, training for improving culture and attitudes, job related training, development and implementing a quality data and reporting system, quality management system implementation, quality audits, inline inspection, incoming material inspection, vendor surveys, process capability studies, design concept reviews, market survey with customer interaction, order clarity meetings, post mortem of any kind to understand the real reasons and attempting to prevent, process control and in-process inspection.

Appraisal costs are the costs incurred for inspection of products, and are termed as bad costs, which verify whether an activity was done properly or not. They just segregate bad products from good, but do not prevent the errors. When most people think about quality control, they normally think about people inspecting a product i.e. quality control inspectors. They inspect fabrics for defects, garment measurements for fitting, print for registration and correct colour, stock from receiving to shipping and think they are doing good, efficient, quality control. In fact, in some businesses 100% inspection is considered the best level of quality control available. One should ask why inspecting a product after the operation is completed (i.e. after making the mistake). Inspection cannot build quality into a product, but will sort out some of the junk. If junk is not produced in the first place, there is no need to pay for the junk and for the inspection. The typical appraisal costs include costs for running an inspection set up, inspection related training, final product inspection, field performance testing, customer satisfaction audit, second-level manager's review of first level management decision, etc.

The field failure costs are results of failures either before are after the materials are dispatched. They are called as ugly costs. The failures before despatch are internal field failures which also include the cost incurred in rectifying an error before the product or service is accepted by the customer. These are incurred because someone did not do the job right. The list includes the in-process scrap and rework or reprocessing, losses due to faulty and spoilt work, examiners records, mending the faults, additional materials, disruption of production, retyping of letter or redoing of reports, fire fighting, repairing the machines, engineering changes, downgrading products, additional inventory to support process yields and rejected lots, overtime, missed schedules, additional men employed for getting quality, etc.

The external field failure costs are incurred by the manufacturer because the customer is supplied with an unacceptable product or service. It is the cost incurred by the company because the appraisal system did not detect all the errors before the product or service was delivered. They include customer rejected service or products, product liability suits, complaint handling, warranty administration, maintenance of field service centers, training of repair personnel, stocking spare parts to support field failure, expenses to rebuild the confidence in customers, replacing defectives or complaint adjustments, reaching agreement and settlement of complaints, office administration, penalties of not meeting delivery dates, failure to meet export arrangements, shipping etc. Besides the direct costs of poor quality incurred by the manufacturer, customers also incur costs, which are termed as indirect external field failure costs. They are the costs to rectify the defects, correspondence with the manufacturer for complaint settlement, business loss due to poor quality received and expenses to repair the loss of reputation.

Studies have shown that the preventive costs, appraisal costs and field failure costs are normally in the ratio 10:30:60 as shown in figure 9.3. Failure costs, as they are the largest, will usually give the largest return for the efforts in reducing them. An effective way of attacking failure costs is through a temporary increase in prevention and appraisal costs; for example, the cost of production and inspection might be reduced by more attention to value engineering, which would to some extent increase prevention costs, and a closer control of the manufacturing process which would increase appraisal costs. Appraisal costs will usually be the next to come under attack. An analysis of all essential quality control operations will often show opportunities for reducing expenditure without reducing effectiveness. For example, statistical sampling techniques may be used as a means of control, indicating trends in performance and assisting to maintain quality. By improving the control of the process, 100% inspection may no longer be necessary. Total costs will be lowest when design staffs are aware of the cost implications of their work. Good design saves cost not only at the design stage but throughout production and testing: products become easier to make "right first time". Good design is needed not only when conceiving the product but also when conceiving systems for production and quality control. After failure and appraisal costs have been reduced by attention to the prevention aspect, it may be possible to reduce prevention costs as well.

In a garment industry, the activities of understanding the order requirement, i.e., order clarity meeting, getting the samples developed and approval from the representative of the customer, auditing of supplier sources, inspection of yarns, fabrics, accessories, inline quality inspections, etc., are classical examples of preventive costs. However, the system of

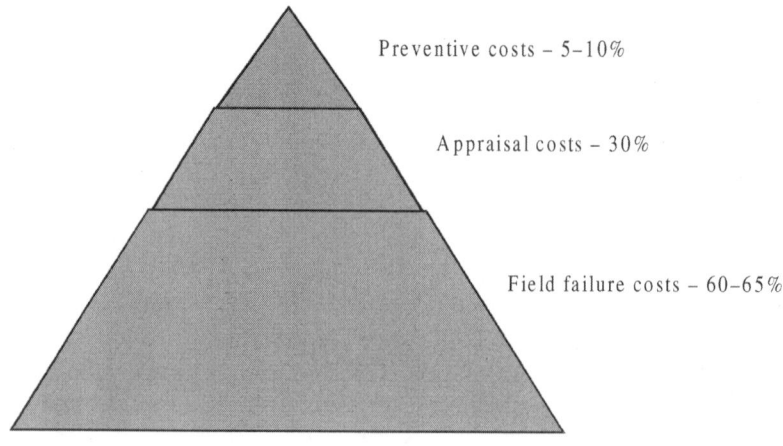

Preventive costs – 5–10%

Appraisal costs – 30%

Field failure costs – 60–65%

9.3 Relative costs of poor quality

recording the customer feedbacks, analyzing them and taking suitable actions, communicating the actions taken down the line and across the organization is found not happening in number of companies, including some of the leading ones. It might be due to the fact that the customers as well as staff are changing very fast, and the quality system concepts have not percolated deep in the minds of top management.

9.5   Controlling the costs

The only purpose of reporting costs is to provoke action. Without action the money spent on deriving and reporting data is a waste. Action is required whenever there is a significant difference between an actual cost and the budget, to discover the reason for the difference and to eliminate it. If cost reports are to be effective in provoking this type of action they must be presented at suitably short intervals quickly following the period they represent in simple, direct, intelligible form to the people who have the authority and knowledge to act effectively. It is often effective if the reports are sent both to the person who is expected to take action and also to his immediate superior.

Effective quality cost control depends upon good cost reporting, which should identify the areas of expense, show actual expenditure compared with that planned, facilitate the comparison of benefits with the price that is being paid, and indicate the causes of excessive costs.

The data should be so organized that further investigations into specific excesses can proceed logically and without the need for re-analysis of basic documents.

Cost reports commonly are in one of three main forms, corresponding to the main divisions of quality costs viz. failure cost report, appraisal cost report and prevention cost report.

The failure cost report reports the level of failure costs like scrap, repairs, test rejections, after-sales service and customer returns etc. The essential data includes the cause of failure, the value lost and the department or process responsible. Supporting data for this report may include reporting point, description of product, part etc., and the responsible machine group or operative. Such supporting data is not needed to include in the reports for executive action but in daily reports for information and action at "shop-floor" level.

The appraisal cost report reflects the cost of operating the quality and reliability surveillance, as compared with budgeted expenditure. The division of account headings may sometimes make it difficult to include the appraisal costs incurred by production operatives carrying out additional operations such as the inspection, testing or grading of piece parts, but such costs can sometimes be derived from a comparison of actual and standard times for the tasks, and included in a separate section of the report. Also it is seen that finding the cost for sample development for a new style, and sample developed for a running style becomes difficult to be distinguished as both are done by the same team and same activities are involved. The appraisal is done for both.

Third type is a prevention cost report. Many functions of the typical business can be interpreted as contributing to prevention costs. It is normally suggested to restrict reports to those areas which are being deliberately varied as part of the overall cost reduction project. The scope of such ad hoc reports can be enlarged to include data from which changes in quality tactics can be planned. Such reports might include an analysis of the effects on profits of changes in the system of setting manufacturing tolerances, the probable cost effects of introducing a vendor rating scheme, recommendations on the most economical points for inspection in a sequence of operations and an investigation into the economics of buying new testing facilities.

Here are ten steps that can be taken as guidance to reduce quality costs;

1. Find out what failure costs are. The cost headings might include as listed below:
    - A - "Prevention costs", i.e. costs of attaining reliability :
      - Quality engineering and testing through pre-production stages; material specifications and design tolerance.
      - Training quality and production personnel in quality attainment.
      - Preparing test specifications and quality standards.
      - Specifying test and inspection equipment.
      - Advising on specifications of the production facilities needed to maintain quality standards.
      - Testing and calibrating inspection and production facilities.
      - Quality administration.
      - Replacement of hand by machine operations.
      - Providing mechanical handling facilities.
      - Providing adequate protective packing.
      - Providing adequate protective storage.
      - Providing bins etc. to protect components during process.
    - B - "Appraisal costs", i.e. costs of maintaining reliability
      - Vendor and incoming inspections.
      - Inspecting and testing products and facilities.
      - Maintaining, re-testing and calibrating inspection and production facilities.

- C - "Failure costs" internal and external
  - Work scrapped: material and labour costs.
  - Sorting out bad work.
  - Reprocessing.
  - Re-inspection and re-testing.
  - Technical and clerical effort spent investigating faults and complaints.
  - Warranty claims, and gratuitous after-sales service.
  - Loss due to sale as second-grade product.
  - Delay in payment by customer - interest on outstanding money.
  - Additional handling charges.
  - Additional transport charges.
  - Additional packing charges

2. Decide, from the size of the preventable failure costs, the scale of extra quality control effort devoted to prevention and appraisal which can be justified.
3. Nominate a senior member of the organisation to have responsibility for quality control who is familiar with and able to lead and train his staff in all aspects of quality control.
4. Obtain a list of actions which can be taken, in the particular circumstances of the organisation, to reduce systematically the failure costs.
5. Evaluate the probable benefit of each action in reducing failure costs.
6. Evaluate the probable cost of each of actions separately.
7. Choose the one or two actions which are seen to offer the probability of largest return for the cost to be incurred.
8. Make the quality controller responsible for seeing that these actions are taken and that the forecast benefits are actually secured. Allot works that are accomplishable and a firm date for completion that is reasonable. Get regular report of progress in cost terms, but don't interfere with the authority that you have delegated to him.
9. As benefits are seen to flow from the first few actions, initiate a few more from the original list and insist that extra possibilities are constantly added to the list so that the process never comes to an end.
10. Find out what the appraisal costs are and, in an exactly similar way, initiate actions designed to reduce specific costs by improving prevention activities.

9.5.1   Essential and avoidable costs and lean concepts

The manufacturing cost if analyzed carefully can be grouped into value-adding costs and non-value adding costs. The value adding costs are essential and cannot be avoided; for e.g. the activities of cutting, stitching, finishing etc., in a garment factory. The non value adding costs can further be analyzed as essentials and avoidable. One cannot avoid all non-value adding costs, as some are essential, like safety measures, meeting the legal requirements etc. Keeping more stocks in the stores or in work places, moving materials from one section to another, keeping additional machines as a stand by either to cope up with the requirements or to prevent production losses in case of breakdowns are examples of non-value adding avoidable costs. Lean concept in manufacturing is a tool for identifying the non-value adding avoidable costs.

The garment industry offers numerous opportunities for improvement using Lean principles. It starts is with a focus on continuous flow or one piece flow. The optimization process focuses on identifying the ideal batch size based on individual manufacturing processes or material handling. In a Lean environment, the ideal batch size is always one. This generally requires work cells are organized by product (rather than process). Under the principle of One Piece Flow, production rates are determined by 'Takt', the rate at which the customers are buying product (the time taken by next operation). Transitioning to a Continuous Flow model requires support and adoption of other related lean initiatives, notably the use of "pull" systems to avoid over-production and schedule-leveling tools. Lean is mainly management of activities of people. Training and involvement of people working on 5 S concepts, maintaining the same team continuously in a line, a foolproof preventive maintenance are essential factors to make the single piece flow logic a success.

9.5.2   Controlling wastes

Waste can be defined as the process or product for which the customer is not ready to pay. Waste minimisation is the process of reducing the amount of waste produced by a person, a system or a society. Waste minimisation is also strongly related to efforts to minimise the use of resource and energy. For the same commercial output, if fewer materials are used, then less waste is produced. Waste minimisation usually requires knowledge of the production process, cradle-to-grave analysis (the tracking of materials from their extraction to their return to earth) and detailed knowledge of the composition of the waste. The main sources of waste vary from place to place. Where the processes are fully automated and costly, a slight inefficiency in labour increases the waste to a great extent.

Waste minimisation often requires investment, which is often compensated by the resulting savings. Waste reduction in one part of the production process may create waste production to another part. Figure 9.4 shows the hierarchy of waste management.

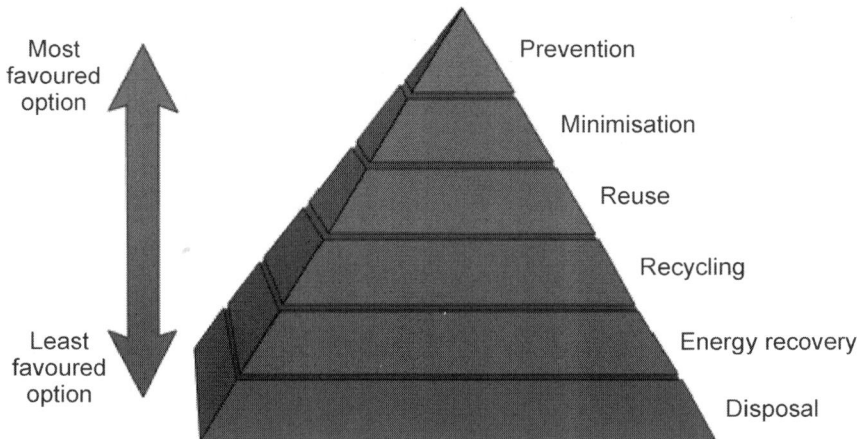

9.4 Hierarchy of waste management

The following is a list of waste minimisation processes:

> *Resource optimisation* – Minimising the amount of waste produced goes hand-in-hand with optimising their use of raw materials. For example, a dressmaker may arrange pattern pieces on a length of fabric in a particular way to enable the garment to be cut out from the smallest area of fabric.
> *Reuse of scrap material* – The introduction of techniques or processes that enable production scrap to immediately be re-incorporated at the beginning of the manufacturing line. For example, in cotton spinning mills any soft wastes generated is returned to the mixing, i.e. beginning of the production line.
> *Improved quality control and process monitoring* – Taking steps to ensure that the number of reject batches is kept to a minimum, by increasing the frequency and the number of points of inspection and by providing adequate training to the operators. For example, inline inspection in a garment sewing batch.
> *Waste exchanges* – Wastes can be exchanged where the waste product of one process becomes the raw material for a second process. Waste exchanges represent another way of reducing waste disposal volumes for waste that cannot be eliminated. Example: The comber noils are given to a lower mixing.

> *Ship to point of use* – Making deliveries of incoming raw materials or components direct to the point where they are assembled or used in the manufacturing process can minimise handling and the use of protective wrappings or enclosures.

> *Product design* – Waste minimisation and resource maximisation for manufactured products can most easily be done at the design stage. Reducing the number of components used in a product or making the product easier to take apart can make it easier to produce with less wastes. In some cases, it may be best not to minimise the volume of materials used to make a product, but reduce the toxicity of the waste created or the environmental impact of the product's use.

> *Fitting the intended use* – A product manufactured for "one off use" should be designed to meet its intended use. This applies especially to non woven diapers, packaging materials, etc., which should only be as durable as necessary to serve their intended purpose.

> *Durability* – Improving product durability, can reduce waste and improves resource optimisation. But in some cases it has a negative environmental impact. If a product is too durable, its replacement with more efficient technology/product is likely to be delayed. View any manufactured product at the end of its useful life as a resource for recycling and reuse rather than waste. Recycling a product is easier if it is made of fewer materials. One has to study between minimising the resources used to make a product and the possibility of reusing or recycling it.

In contrast to waste minimisation, waste management focuses on processing waste after it is created, concentrating on re-use, recycling, composting and waste-to-energy conversion. In industry, waste is generally reduced by using more efficient manufacturing processes and better materials. The application of waste minimisation approaches has led to the development of innovative and commercially successful replacement products. Waste minimisation has proven benefits to industry and the wider environment as it reduces raw material costs, the cost of transport and processing raw materials and the finished product and the waste disposal cost to other parties (including collection, transport, processing and disposal). Proper collection of wastes, not allowing them to get mixed with other wastes, disposing it off at the earliest are very important if we need to get better realization from the wastes generated.

Problem solving and change management

10.1 What is a problem?

The management expects the supervisors to solve the problems occurring on day to day basis. We discuss on problems in the industry. Each one has a problem or a number of problems, whereas all might not be having same problem. Some problems of mine are "not at all a problem" to others. Then what is a problem? There are different definitions like a doubtful or difficult matter requiring a solution or something hard to understand or to deal with. It can be defined well as an undesirable result of a job, as told by Dr. Juran. The solution of a problem is to improve the poor result to a reasonable level. If we solve our problem, we can move forward. If we leave it, it shall grow.

Whether something is a problem or not is depending on the person seeing it. For example someone may say that their ring frames are old and cannot run at the same speed as that of latest ring frames and hence they are unable to compete. However, someone may take it as an opportunity and claim that the yarn spun has higher elongation at break and hence works well in weaving and knitting, or use for certain types of fancy yarns or special yarns where slow speed working is needed.

A path to success has number of hurdles. If you like going in a path, where there are no problems, it means you have taken the easiest route in which all are moving. In such a case, you are one among them and cannot be a winner. Don't be very happy if you are moving at a high speed without any hurdles. Remember, a road without hurdles can take you down and not up. When you have to go up, you have to come across hurdles one, who is afraid of facing problems, can never be a winner. So be ready to face the problems, and not try to avoid them or postpone your actions. If you are hesitating to move forward, someone shall overtake you, and you cannot come out as a winner.

10.2 Roots of a problem

I have a tree in my yard, which is not wanted (Fig. 10.1). It is a problem to me, as it sheds lot of leaves, and house is getting cracked because of its

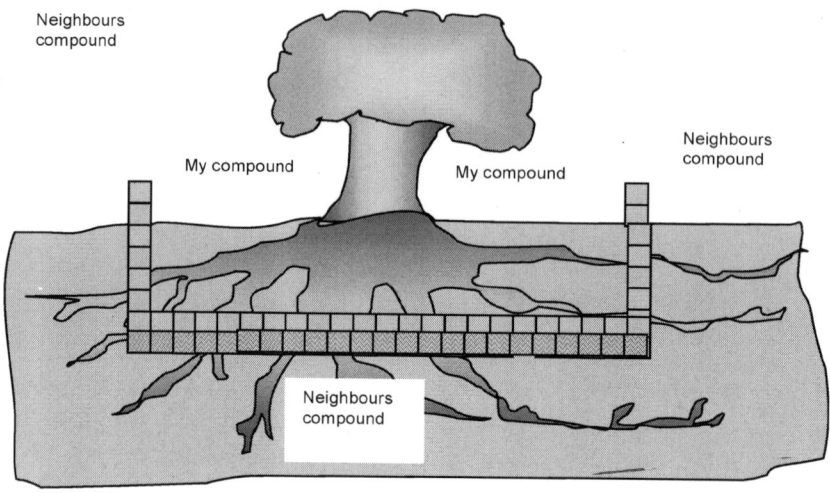

Neighbours compound

Neighbours compound

My compound

My compound

Neighbours compound

10.1 The Tree in my yard.

roots. I have tried to cut it off, but still it is growing. I am not putting any water, but it is managing by taking water from the neighbours. What can I do? How to solve the problem?

When Dr. Juran told that around 80% of the problems were management related, people laughed at him and branded him as mad. Let us see my example. I have a tree which is a problem to me. Who is responsible for that? I did not remove it when it was a small plant. I allowed it to grow. The roots spread. I am not putting water, but it is managing. Is my neighbours are responsible? No. They are not aware that roots are there in their yard. Why the problem came? The land and soil in my yard is suitable for that tree to grow. My systems are favourable for the problem to grow. I should have reinforced my land by putting stones or cement, so that a plant cannot grow. Now what I have to do? I should take the neighbours into confidence, and dig the roots out from their yard. Once all roots are removed, I have to put stone pavement to prevent a second plant from growing in my yard. This holds well for all the problems and for all organizations.

The analysis made by American association for safety and health revealed that behind every fatal accident there were 30 near misses and 300 unsafe conditions (Fig. 10.2). To prevent one fatal accident one has to concentrate for proactively eliminating 300 unsafe conditions. The unsafe condition is easy to attend like a problem in the seed stage. But people ignore the signal given by the spread of unwanted seeds. The near misses are like a problem in the seedling stage. It is possible to remove the seedling with slight effort, but people do not find time to do that. Finally when it

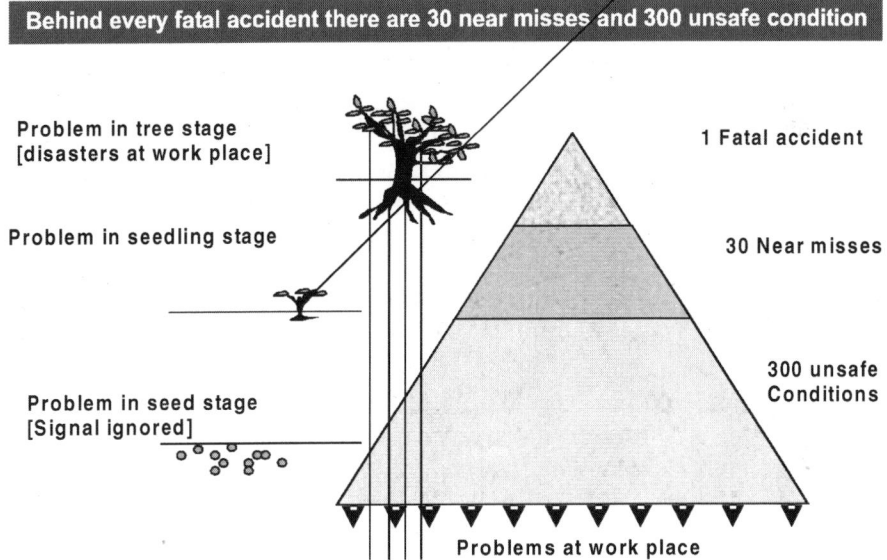

Behind every fatal accident there are 30 near misses and 300 unsafe condition

Problem in tree stage
[disasters at work place]   1 Fatal accident

Problem in seedling stage   30 Near misses

                                                             300 unsafe
Problem in seed stage   Conditions
[Signal ignored]

                    Problems at work place

10.2 Theory of 1 - 30 - 300.

becomes a tree, it is not possible to remove, or not viable to remove. People learn living with the problem or become a victim of it.

10.3   Seven steps for problem solving

We normally see the symptoms of a problem and not the problem itself. For example, if we say breakages are high in a particular ring frame, the real problem lies with that machine, may be a worn out part or vibrations or wrong combination or meshing of gears and so on. We need to identify the reason. To solve a problem we need to identify the root cause and attack. For finding the root cause, we need to go on asking "Why" a number of times. This technique is called as "Five Why Technique". We need to analyze the boundary of problem. We may have to brain storm with the concerned and list all the possible reasons, and make a cause and effect analysis as to which of the reasons are prevalent as on date. Then analyze all the reasons by careful observation.

Seven steps for problem solving were designed by Dr. Juran and Prof. Ishikawa, which are relevant for almost all situations. They are as follows.

1. Problem identification – Define the problem clearly
2. Observation – Recognition of the features of the problem.
3. Analysis – Finding out the main causes.
4. Action – Action to eliminate the causes
5. Check – Confirmation of the effectiveness of the action.

6. Standardization – Permanent elimination of the causes.
7. Conclusion – Review of the activities and planning for future work.

Problem identification – The first step is defining the problem, highlighting its importance and the need to solve it. It is suggested to explain the background of the problem and the course of actions taken so far, the loss because of it, the extent to which it is required to be solved during the present task, and the budget sanctioned for solving this. Use of as much data as possible is essential to identify the most important problem. One should have strong reason for selecting a problem for solving. The circumstance in which the problem gets priority is to be identified and highlighted. The undesirable results and losses due to the poor performance are to be expressed in concrete terms. If the degree of importance is high and widely understood by many people, the problem will be dealt with seriously. The basis on which the target values are set in the theme and its importance are indicated. When the theme includes many kinds of problems, then they are to be divided into sub themes for effective handling of the problem.

Observation – This step consists of investigation of the specific features of the problem from a wide range of different view points like time, place, type and symptom. As the intensity and effect of problem depends on various factors, the investigation must be done from different points of view to discover variation in results. One needs to go to site and collect necessary information that is not put in the data form. The objective of this step is to discover the factors that are responsible for causing the problem.

Analysis – The third step is to find the main cause of the problem. The activities include setting up of hypothesis, verification of links, deleting non-relevant elements and testing the hypothesis to identify the real root of the problem. Use of cause and effect diagram is made to ensure collection of all possible knowledge concerning possible causes. The information obtained in the observation step is used to delete the elements which are not clearly relevant. New plans are devised and data collected to ascertain the effect of those elements on the problem. The data is validated by small experiments, and some time the poor result is intentionally reproduced.

Action – The fourth step, 'Action', is to eliminate the causes. A strict distinction must be made between actions taken to cure phenomena (immediate remedy) and to eliminate casual factors (preventing recurrence). The ideal way of solving is to prevent the problem from recurring by adopting remedies to eliminate the cause. While taking action care should be taken to ensure that they do not produce other problems. If it is inevitable, devise remedies for the side effects. The action has to be thoroughly evaluated and judged from a wide range of view points as possible. It is advisable to conduct trials and check. Devise a number of different proposals for action; examine the advantages and disadvantages of each and select those which the people involved agree to. An important practical point in selecting actions is ensuring the active cooperation of all those involved.

Check – The next step is checking to ensure that the problem is prevented from recurring. This involves comparing the results before and after the implementation of the solution. The data should be collected in the same format as it was done for analyzing the problem and comparisons made. It is also essential to relate the situations before and after for understanding. If the undesirable results continue even after actions have been taken, the problem solving has failed. We need to go back to the observation step and start again.

Standardization – The step 'Standardization' is to eliminate the cause of the problem permanently, by devising the procedures and documenting them. Without documenting the procedures and standardizing, the actions taken to solve the problem will gradually revert to the old ways and lead to recurrence of the problem, and also it is likely to revert when new people are on work. Standardization will not be achieved simply by documents. It must become a part of the thoughts and habits of all. Education and training to all involved along with assigning responsibilities is an important part of this step.

Conclusion – The last step is reviewing the problem solving procedure and planning future work. The activities involved are summing up the problem remaining, planning the work for solving the remaining part and verifying what went well and badly. One should realize that a problem is never perfectly solved and an ideal situation almost never exists. It is not good to aim for perfection or to continue the same activities on the same theme for too long. When the original time limit is reached, delimiting the activities is important. Even if the target is not reached, a list should be made of how far the activities have progressed and what has not been attained yet. It might be worthwhile to live with a problem rather than eradicating it fully; which depends on the nature of the problem remaining, it's after effects, the steps and cost needed to eradicate it fully.

A systematic approach as mentioned in seven steps above reduces the number of problems and helps in moving towards zero.

10.4   Use of QC tools

The most important step in problem solving is the identification of the real problem and its root. We need some tools for analysis and diagnosing the problem. Those tools are popular as QC Tools. Seven QC tools were recognized during the evolution of TQM concepts in Japan. They are data collection, check sheets, stratification, brainstorming, cause and effect diagram, pareto analysis and scatter diagram. As the time passed number of other tools were recognized and added in the list. They include histograms, force field analysis, critical activity chart, flow chart, concentration diagram, run chart and control charts, spectrograms, etc. To understand the impact of potential problems, failure mode effect analysis was developed for use at designing stage itself, which is supported by quality function deployment for optimizing the process parameters. For implementing the actions management tools were developed, which include affinity diagram, relation diagram, tree diagram, matrix data analysis, matrix diagram, process decision programme chart, arrow diagram etc. Let us discuss some of the QC tools.

10.4.1 Data collection

Data is the building block on which fact based decisions are made. The collections of facts and figures which can give a clear picture of a required work situation are called data. Data collection is the most important factor influencing the success of a problem identification process, which is the first step in any of the improvement projects. The data would form a sound basis for decision making and corrective action. The method of planning, organizing and auditing the process of data collection are key factors, which can 'make' or 'break' any improvement effort.

The primary purpose of collecting data is to answer questions, which may come from opinions during different stages of problem solving and decision making. Accordingly data is collected. It is necessary to verify the reliability and correctness of data; its relevance to the problem and ability to reveal the facts.

Data should be either measurable or countable to make analysis and study the trends. There are some data, which can neither be counted nor measured like smell, taste, fastness properties of dyed material, yarn or fabric appearance, feel of a fabric etc. It is needed to convert them into some measurable terms. Information can be obtained by careful observation of facts and analysis of data. The data can be obtained by referring to past records, actual measurements, enumeration, sampling, controlled experiments, surveys etc. In a number of cases, data required may be available in some form, and we need to put them in a required form to facilitate analysis. Which data and how much to be collected depends on the nature of problem. The following 10 steps can be used as a guide to design a data collection system.

1. Formulate questions which can lead to the root cause.
2. Consider appropriate data and analysis tools.

3. Define a comprehensive collection plan.
4. Anticipate bias in collecting data, and take measures to avoid or minimize.
5. Understand the data collector and their environment.
6. Design a simple data collection form.
7. Prepare the instructions for use.
8. Test the forms and instructions.
9. Train the data collectors to be focused on the problem.
10. Audit the collection process and validate the results.

One need to ensure that right type of instrument is used for data collection that is calibrated and maintained. Proper definition for classification should be made for enumerating data. Factors which can influence data are also to be recorded. Data should be recorded honestly and nothing should be left out.

10.4.2 Brainstorming

Brainstorming is used to help a group to create as many ideas as possible in a short time. Normally single person cannot have the complete experience or knowledge of a situation, so it is suggested to involve all concerned from various sections where the roots of the problem are spread. The subject is to be made clear and specific to the participants as it helps to focus their thoughts and ideas. Each member tells one reason at a time in rotation, and when he is not ready, shall say 'pass' and allow the next person to tell. There shall be no discussions while the points are being told, and no one will laugh or comment on the points. There is no need of giving any explanation or justification while telling the points. All points shall be recorded on a black board or a flip chart to avoid repetition. This is a good group education technique, eliminates bias to some extent, and brings a feeling of oneness in the team as the participants sit together in sharing their experiences.

10.4.3 Flow charts – process mapping

To analyze a problem and finding solution, it is necessary to understand the process. Mapping of the process and preparing a flowchart (Fig. 10.3) showing sequentially the inputs, activities and processes, checking done and the controls exercised, the feedback loops, the decision points, intermediate and final outputs help in understanding the process. The flowchart is self explaining and does not give any interpretation by itself, however, when the ideal flowchart is compared with the actual, it shows points of deviations. A new process or plan can be tested for its logical consistency by following all paths of the chart.

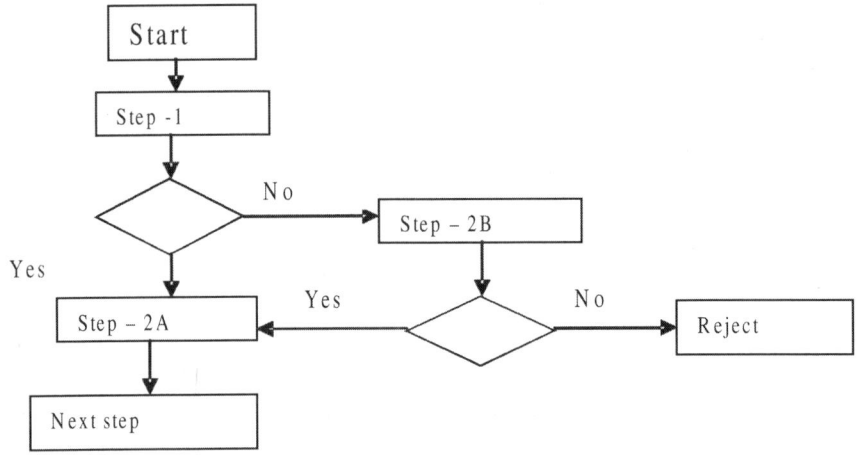

10.3 Typical flow chart.

When we get a problem, it is essential to analyze the path in which the work took place, and verify the reports and the actions mentioned. Identify the possible reason for this problem to remain unnoticed during the process. If we have a habit of documenting all the events that took place while carrying out an activity, we shall be able to identify the problem in its root itself, and does not allow it to grow. In majority of the cases, we fail to identify the problem in the beginning because of lack of documentation.

10.4.4 Critical activity chart

After understanding a flowchart, information is gathered on each step to understand its criticality in process. The critical activity chart is a tool for systematically gathering and analyzing information about job process or operations, concentrating mainly on inputs, outputs and basic processes, rather than specific interlinked sequence of process steps. This is used to understand the work process and define its boundary in problem identification stage, help brainstorming in work place and to identify major causes for dissatisfaction of internal customer, to identify the major areas of output and their internal customers and asses the extent of their satisfaction. To prepare a critical activity chart, the major work activities are identified and for each activity one chart is prepared. The next step is identifying the inputs, outputs, tasks and the customers. Then problems or deviations in output, input and processes are identified. Problems can be clarified by discussing the contents with all concerned. It is a tool for common understanding. A good critical activity chart (Fig. 10.4) helps in highlighting the workplace activities and if analyzed systematically makes Brainstorming faster and more useful.

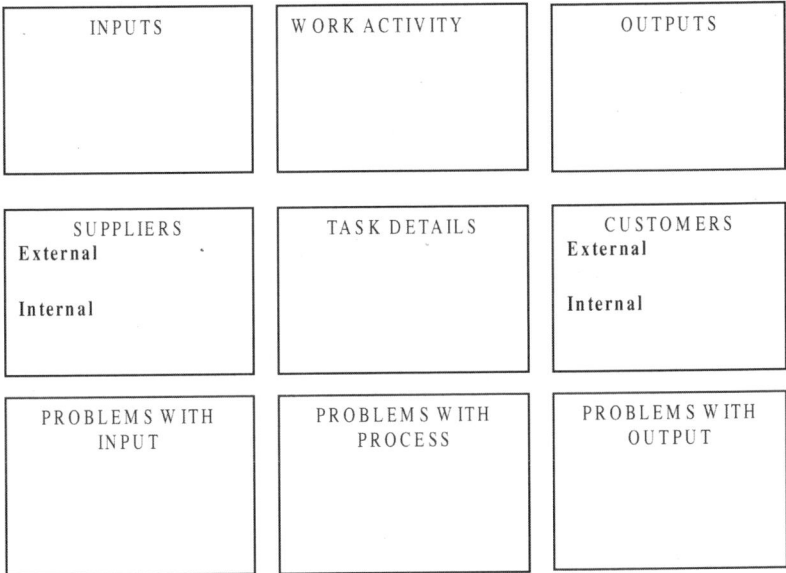

INPUTS	WORK ACTIVITY	OUTPUTS
SUPPLIERS External Internal	TASK DETAILS	CUSTOMERS External Internal
PROBLEMS WITH INPUT	PROBLEMS WITH PROCESS	PROBLEMS WITH OUTPUT

10.4 Critical activity chart.

10.4.5 Boundary analysis

A boundary analysis illustrates the relationships that exist within the processes. It details all interactions among processes, customer and suppliers. First the starting and ending points of the process are defined. Then details are prepared for each process.

Customer input → perceptions, requirements, complaints, expectations

Process output to customer → information, deliverables (products/
services)

Supplier input → information, deliverables

Process output to supplier → perceptions, requirements, complaints, expectations

Resources → equipment, people, budget, procedure and training.

10.4.6 Check sheets

Check sheet is a well thought out format for collecting and compiling data for events as they happen which makes it easier for subsequent analysis. It may be a form, format or table. The check sheets are useful in the areas to understand the past and present status of the problem situation, to stratify the data as they are collected, to understand the change through the passage of time trend, to analyze the data as they are collected, to determine the details of defects, to determine the source of defects, to inspect machines or equipments and to verify the operating procedure.

The check sheets are of three groups, viz. check sheets for recording data and making surveys, inspection and validation check sheet and check drawings. The check drawings are helpful in locating the exact location of defects to identify problem area. Concentration diagram is a type of check drawing (Fig. 10.5).

The preparation of checklist has several steps. It starts with the clarification of the objectives by clearly stating the event or issue being observed and data pertaining to it to be collected. Depending on the type of the problem one need to decide on when and where the data is to be collected and type of check sheet. Depending on the role of each item in the problem the items to be checked are decided. First a trial check sheet is prepared to ensure its suitability for collecting the data. The check sheet should include; the title, object/items to be checked, checking method, date and time of check, the checker/observer, the location and the summary of conclusions. While recording the observations, simple note using symbols can be made so that maximum information can be gathered in one stroke. The information is to be tallied for their completeness. Completely filled check sheet offers clearly visible data for the event and is self explanatory.

10.4.7 Concentration diagram

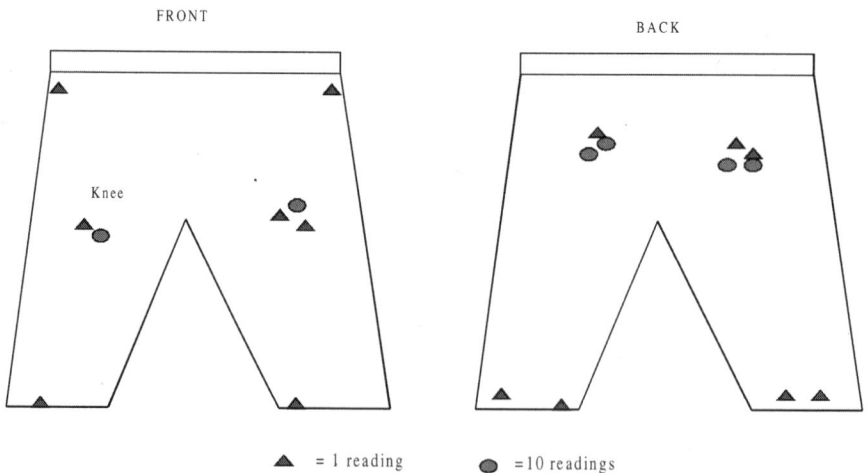

10.5 Concentration diagram - wearing out of trousers.

Concentration diagram is a special check sheet to record data about frequency, type and location of events (defects or errors) on the picture or schematic drawings which is easily understood and visualized. It is used when visual picture or layout of location of event is more clearly understood than possible description, the possible locations are many and proper classification and expression by words are difficult, and when standard check sheets becomes difficult to understand for data collection in remote locations like the exact point of defect. The concentration diagrams when completely filled show the frequency and location of the event. The form is self explanatory, as it indicates the location of the diagram. This can be further analyzed by using other QC Tools.

10.4.8 Stratification

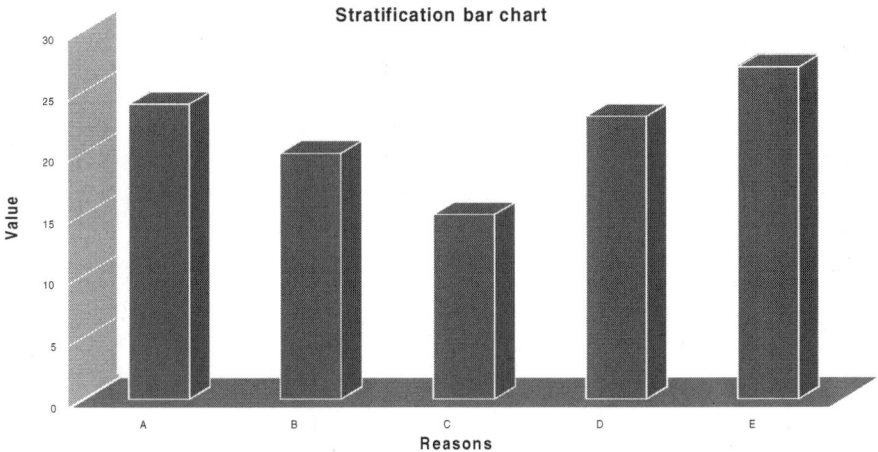

10.6 Stratification.

Stratification (Fig. 10.6) is a statistical technique of breaking down values and numbers into meaningful categories or classification to focus corrective action or to identify true causes. This is used to identify the category which contributes to the problems being tackled. Graphs are among the simplest and best techniques to analyze and display data for easy communication. Stratified data is normally displayed in bar charts, which show comparative characteristics by the length of the bar.

Selection of stratification variables is essential for planning the variables and collecting data, rather than going on adding the variables as and when some information is obtained. The cost and time for collecting additional identifying variables in the initial times is lesser compared to the cost of new collection effort. The next step is to establish category to each variable, which is a value or a range of values of a stratification variable. Then data is collected pertaining to those variables. The data obtained is sorted and grouped into stratified variables. If the first attempt of stratification does not reveal any significant pattern, data is collected again to find out the effect of other variables.

Instead of using a bar chart, use of a run chart is adopted for analyzing the system related problems. Let us take an example of a process and the time taken for doing different processes as shown below.

Time required for the process - minutes						
Process	**A**	**B**	**C**	**D**	**E**	**F**
1-Mar-07	20	43	25	32	25	12
2-Mar-07	21	40	25	35	26	14
3-Mar-07	20	35	25	31	24	16
4-Mar-07	21	43	26	30	25	17
5-Mar-07	23	30	25	28	26	13
6-Mar-07	21	26	26	42	27	15
7-Mar-07	20	40	25	35	25	16

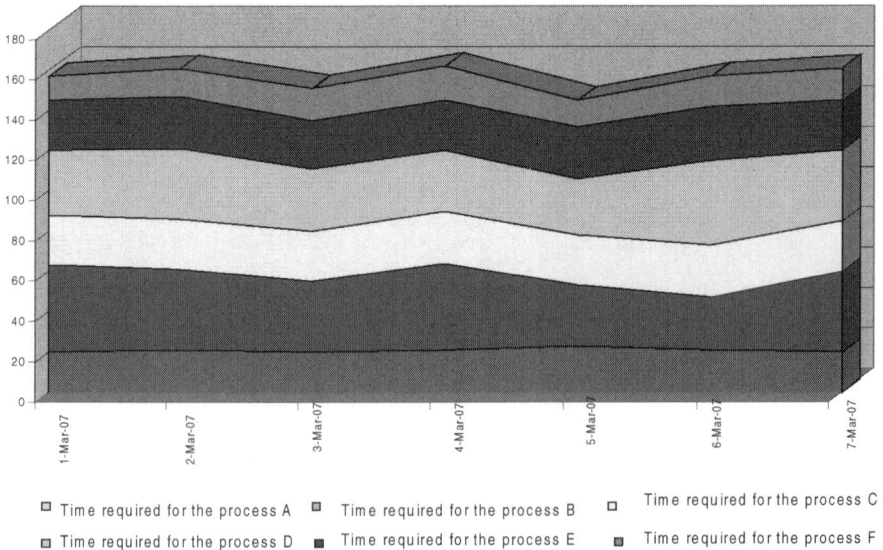

☐ Time required for the process A   ☐ Time required for the process B   ☐ Time required for the process C
☐ Time required for the process D   ■ Time required for the process E   ■ Time required for the process F

10.7 Stratified run chart.

In a job numbers of steps are involved with each having its own importance and criticality. Time taken is different for each element of the job. When we analyze the time taken for each element of a job, the normal practice is to collect data element-wise and project in a Bar chart, where either the average time or total time is shown. However, this cannot identify whether there is a real scope for improvement or standardization or a technological change is required. When we draw a stratified run chart as shown in fig 10.7 we can identify the process which have wide variations and the processes that are consistent. The processes with variations can be improved by standardization, whereas if a process is consistent, improvement can be achieved by technological change. Stratified run charts can be used for reducing the wastes, improving efficiency, implementing Lean systems, delays in maintenance activities, etc.

10.4.9 Run-charts and control charts

Run-charts [Fig 10.8] display the trend of changes of a character over a period of time. The X-axis always refers the time and Y-axis indicates the character under observation.

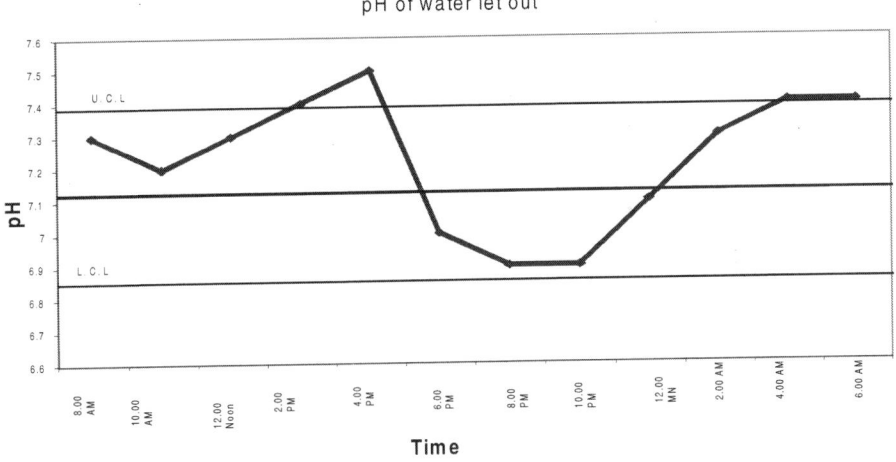

pH of water let out

10.8 Control chart.

This is a specialized graph, which uses connected lines instead of bars to illustrate data. A run-chart with statistically derived control limits is a control chart. This is used to highlight the variations of a characteristic over a period of time and seek explanation for changes, to study the growing or declining trend of the average, to highlight significant improvements in performance after implementing counter measures and to verify the effectiveness of the measures taken for improvement.

To make a run chart, time intervals are marked in the horizontal axis. The numeric scale must move in regular intervals. The vertical scale is marked considering the expected range of variation. A typical run chart is shown in the figure for pH of water let out from a process house in fig 10.8

The run-charts are interpreted by identifying points in time when the characteristic changes significantly. If it is compared with other possible changes at the same time, it offers clues for causes. A common trend may be increasing or decreasing. The action is to be taken if the reading goes out of limits or continuously remain in the border near any of the control limits.

10.4.10 Cause-and-effect diagram

Cause-and-effect diagram is a representation of the systematic relationship between the event under investigation and all possible causes influencing (Fig. 10.9). It is also a documentation of group thinking process to investigate the root cause of the event. As it looks like a skeleton of a fish, is called as fish bone diagram and also as ishikawa diagram in the name of its founder. This is used to investigate the cause and effect and help stratification for collection of data to confirm relationship and evolve counter measures. The steps involved are defining clearly the problem or effect or event for which the cause is to be identified. A horizontal line with an arrow at the right hand end and a box in front of it is drawn. Problem statement is written in the effect box. The next step is identifying the causes in major categories. Brainstorming is normally used to identify the possible major causes. After identifying a primary cause, the team shall go in deep and identify as many secondary or tertiary causes as possible in each of the primary causes. Each of the major causes is placed in a box horizontal to the first line and connected to that line at an inclination of approximately 70°. After identifying the major causes, the root causes are investigated by adopting root cause analysis techniques. The logical validity is checked for all causes identified considering the present scenario.

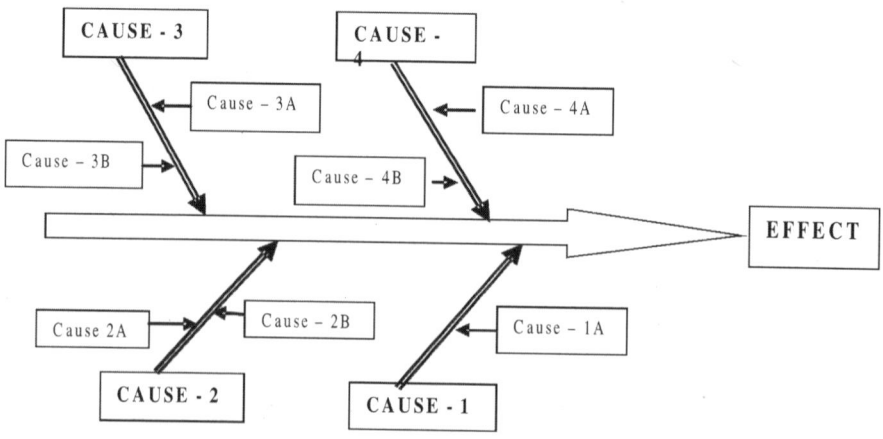

10.9 Cause and effect diagram.

It is important to understand the potential pit falls while using cause and effect diagram. It should not be treated as a substitute for data. It should be drawn only after preliminary data has been collected to narrow down the focus of a problem. One should not limit himself just to those theories which are in the diagram.

10.4.11 Pareto analysis

A pareto diagram (Fig. 10.10) is a special form of vertical bar graph which helps in identifying "vital few" from the "useful many". Its concept was given by Mr. Wilfred Pareto, an economist from Italy, and was developed as a QC Tool by Prof. J. M. Juran. The principle involved is that very few causes contribute for maximum effect; whereas a number of other factors contribute only for a small effect. It is used for setting priority while selecting the problem, and for identifying the most important root causes contributing substantially to the problem.

The impact of each factor is worked as a percent of total impact, and the factors are arranged in a descending order. A bar chart is prepared. In addition to a line graph is prepared for the cumulative impacts worked starting from the highest contributing factor.

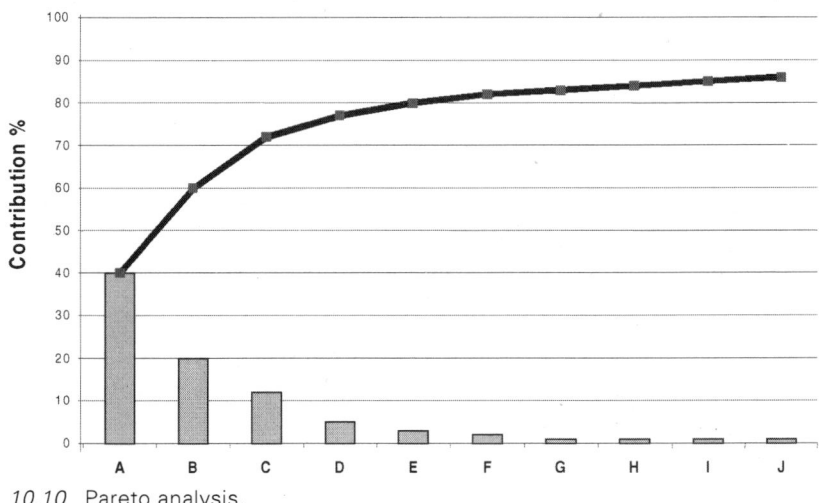

10.10 Pareto analysis.

10.4.12 Histogram

A histogram is a special type of bar chart to show the distribution or spread of the observed characteristics, which enables one to see patterns that are difficult to see in a simple table of numbers. It is a visual presentation of range, magnitude, central tendency and the spread. Histogram was first developed by French statistician A. M. Guerry in 1833. The histogram helps to identify whether a spread is normal or not, the surprises in the natural distribution which can lead to causes or counter measures, confirm the results of a process improvement and to obtain clues for stratification. The common histogram patterns are normal or bell shaped distribution, double peaked distribution, plateau distribution, comb distribution, skewed distribution, truncated distribution, and isolated peaked distribution or island distribution.

10.4.13 Scatter diagrams

Scatter diagram is a simple graphic presentation of the relationship between two variables, which relates to cause and its effect. It is one of the oldest applications of graph and was developed by Dr. Burton in London in 1794. In 1832, Mr. J. F. W Horshel fitted a curve to the scatter diagram. The points marked on a scatter diagram forms a pattern which indicate the degree and nature of relationships, which is statistically known as correlation.

Scatter diagrams are used to verify if there is any relationship between cause and effects with facts and to estimate the strength and nature of the relationship between two sets of data. The concept is that there is always a relation between a cause and effect, but it would be difficult to state them in precise mathematical terms. It is easier to see the relationship in a scatter diagram than in a simple table of numbers. The effective problem solving is possible only when we discover and test the true relationship between a cause and its effect. Normally, the suspected cause is taken in X axis and the effect in Y axis. Points are plotted for each reading of cause and effect. The pattern generated by the cluster of points gives clue to the possible relationship.

The scatter diagrams indicate the relation as strong positive, strong negative, weak positive, weak negative and no relation. There may be some complicated relations, where the value of Y increases as X increases to a certain extent, and then it might change its direction. The relation need not be linear all the time. Depending on the degree of relationship, further analysis or verification can be carried out. Also quantification of degree of relationship can be carried out by working out coefficient of correlation.

10.4.14 Force field analysis

In the improvement process, if it is to be successful, some change has to take place. There are some hindrances for the change and some elements support the change. Force field analysis is a technique developed by Kurt

Lewin to identify elements which resist the change (Hinder) and which are pushing for change (Aid). This helps in developing the implementing strategies for a change by carefully working with the factors which favours or hinder the process. This is used for chalking out possible implementation strategy for an improvement, to forecast and assess the problems likely to occur from hindering factors while implementing a change, and to develop counter measures to minimize the impact of hindering factors during successful implementation.

If the helping factors are more powerful, the change takes place and implementation of counter measure is successful. Normally, each hindering factor has a counter factor which can help. Force field analysis should be analyzed to find these couples (Fig. 10.11). If this is not obvious, it is possible to generate additional 'drivers' to facilitate implementation. Then if helping factors are nurtured, implementation has better chances of success of success.

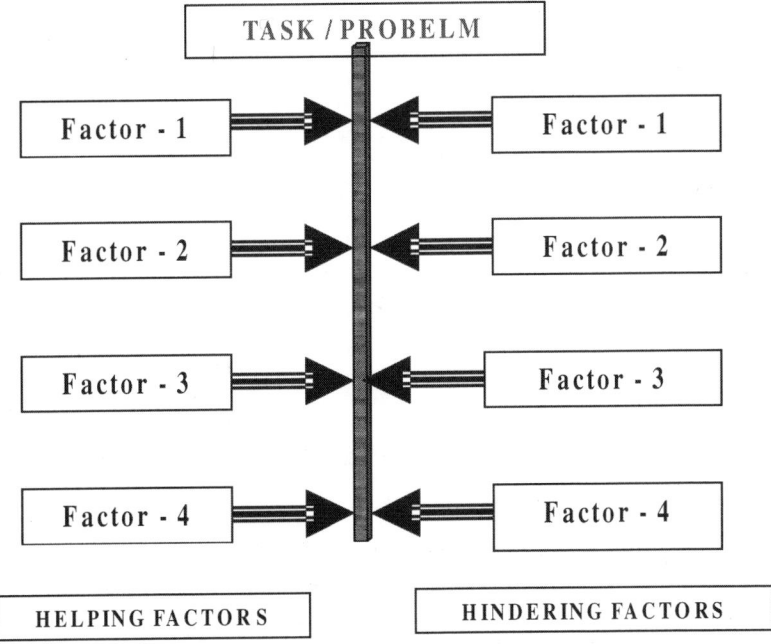

10.11 Force field analysis.

10.4.15 Spectrograms

Spectrogram is a QC tool (Fig. 10.12) being used in textile mills to locate the source of fault in a yarn, filament, rove, sliver or any such continuous strand, which are produced by using rotating rollers. It highlights the defects occurring in a regular frequency. By carefully studying the gearing diagram and working out backwards, it is possible to pinpoint the source of defect.

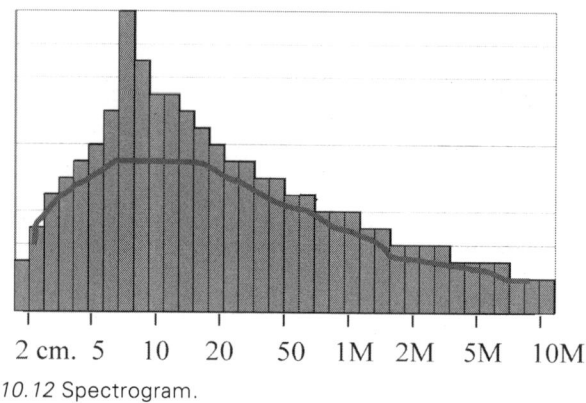

2 cm. 5   10   20   50   1M 2M   5M   10M

10.12 Spectrogram.

The curved line shown in the diagram is an ideal curve drawn considering the effective fibre length, whereas the actual values, i.e. the frequency of a particular type or size repeating are shown as a histogram with blue bars, and is generated by the evenness tester. If the height is more than the ideal curve, then it needs to be corrected. However, the readings which are more than twice the height of ideal curve are considered as significant. In the diagram 10.12 the highest peak is at 8 cm, indicating a defect occurring at every 8 cm of the product. If the draw rollers have a diameter of 25mm, then for every one revolution we get 25 x 3.14 = 7.9 cm. It means there is a problem in the draw roller. Similarly, we can work out the source of defect by understanding the gearing and working out the length of material produced for one revolution of each roller or gear.

10.5   Diagnosis and remedial journey

Good diagnosis is half cure. Diagnosis is carrying out the detailed analysis needed to identify the root cause of a problem. The process starts from the symptom. In the majority of cases a chronic problem shows its symptom in one department or area, while the cause lies in a different section or department. In such cases cross functional teams shall come to help.

Direction and diagnosis are the two processes in a diagnostic journey. The direction provides the project definition and various theories that are to be tested. The theories are outcome of the logical thinking, previous experiences and the knowledge of the members involved in diagnosis. When theories are tested, a number of them turn out to be invalid, whereas some shall emerge as valid and become the basis for subsequent remedial action.

The steps involved in diagnosis are analyzing the symptoms, translating theories into data requirement, testing the theories, analyzing and summarizing the results. The persons involved in diagnosis must have time, diagnostic skills, objectivity and a factual approach. The time is needed to carryout numerous tasks, viz. precisely defining each symptom in order of frequency, applying Pareto principle, designing a plan of data collection and analysis, carrying out data collection, conducting data analysis, summarizing and presenting the results. The diagnostic skills are concerned with the scientific testing of theories; designing the plan for data collection, collecting the data without bias, and bringing the meaning out of the resulting data. Chronic problems continuing for years give raise to long standing biases and it is difficult to convince the people that a problem existed. The factual approach helps in eliminating biases. Hence data is the lighthouse for diagnostic journey. In case of diagnostic skills, there is usually a remarkable difference in the skill level demanded for break through compared to that demanded for trouble shooting. In trouble shooting, the objective is to restore status quo, whereas in break through, it is to reach a level of performance never attained before. It is a voyage of discovery, and the key question is to identify the variables that stand between us and getting rid of this chronic COPQ. In order to get there, we need a map or a procedure which lays out the route. The diagnostic journey consists of the phases; analyzing symptoms, formulating theories, testing theories and identifying the root causes.

One of the important concepts for effecting improvement is the concept of controllability. Identify if errors are worker controllable or management controllable. The analyses of errors in most studies indicate that over 80% are management controllable and another 20% are worker controllable. An error is worker controllable if workers have the means of knowing what they are supposed to do, means for doing and have means available for regulating the performance. The supervisor has a role in communicating this to workers in a language understandable by them.

After identifying or understanding the symptoms, the next step shall be formulating and arranging theories. The progress in diagnosis is made theory by theory, by affirming or denying the validity of these theories about the causes. The theories are to be identified by brainstorming, and it is necessary to record all theories without challenging them. The best source of theories is the men on job. They might have even experimented to get rid of the problem. As the list of theories grows, it is useful to arrange it in a way which shows the interrelation of the various theories. Ishikawa diagram can be used to identify the relation and interaction of theories.

After listing the theories, next step is to test the theories. Which theory should be tested is decided by the experience of managers. Some of the theories can be tested readily using available data, whereas some might involve pains-taking experiments, which if designed improperly are likely to end in frustration. There are four broad methods of approach for testing theories by using existing data, by using current operations data, by cutting new windows and by designing and conducting experiments.

By testing various theories we get the resultant data that point to the possible cause. Once the cause is known, it is relatively easy to arrive at a consensus on effective remedial action. The journey ends when the remedy has been demonstrated capable of producing improved results under operating conditions.

To ensure that the chosen remedy is optimal from the company's prospective, and it is successfully implemented, the following guidelines are to be used:

- Choose from alternative
- Anticipate resistance to change
- Implement remedial action
- Establish control at the improved level
- Suggest where the remedy can be replicated
- Feedback repetitive problems to product and process planners

There may be alternative solutions for a problem, and one needs to choose from the alternatives by considering the cost involved, ease for implementation, acceptability to implementers and the probable side effects etc. The necessity of making conciliation occurs every time an idea or innovation needs to be assessed. Suppose a new system has been thought up to meet a problem or situation which needs to be improved, the question before getting into it shall be a detailed planning of such a system and a confirmation that it pays. This question needs answering before much time, effort and money are spent.

Once the remedy is accepted, it should be implemented by the users, but it is observed that the team encounters delaying tactics or outright rejection of the remedy. The source of resistance varies, which may be a manager, a supervisor, the workforce or the union. Always there are reasons to resist, and surprisingly in a number of cases, such reasons for rejection are advanced by the very people whom the change is intended to benefit.

We normally deal with two changes, viz. a technological change and the consequences of the technological change. The social consequence is the trouble maker, which is a sort of uninvited guest and rides on the back of any technological change. A systematic approach of taking the people into confidence, starting small, transparency and without surprises, choosing the right time, and moving with the culture of the society are helpful in a remedial journey with least resistance. Implementing remedial actions includes providing a capable process to the operating forces which can hold the gains, establishing operation standards and procedures to serve as a basis for training, control and audit, providing training to use the procedures and to meet the standards and developing a system of control for detection and correction of out-of-control conditions. Establishing control at improved level is very essential to hold the gains. This might be done by process or operations audit and financial controls.

Replication of remedies for a similar quality problem elsewhere by suitable communication is essential to get the full advantage of the remedial journey. The design review should verify the remedies taken for various problems earlier, so the process can be designed properly with lesser problems. A proper diagnosis and remedial action can help in solving the problems from their root and lead to permanent solution and improvement. Hence the supervisors can be winners by following the steps religiously.

10.6   Manage the change

Identifying a problem and finding a solution might be an easy task compared to implementing what was decided. One has to face lot of challenges while implementation. We shall be bringing a change while implementing a remedy for a problem. We might bring change in people, technology, thinking, policy, procedures and systems, environment, which all tend to affect the equilibrium. Modification in the way of performing certain jobs, change in rules and procedures, adoption of a new technology, changes in organization structure etc., affect the internal equilibrium, whereas change in market situation, government rules and regulation, political scene, economic scenes etc., affect the external equilibrium. One has to deal with changes to reduce the tension by understanding and altering the forces and making people participative and committed. In order to bring change effectively, one needs to understand the following.

- The personal views and experiences of individuals
- Possibilities for resistance to change
- The possible reasons for resistance to change
- Reasons for going along
- How change affects the individual
- How to help people move smooth through the road to change
- Typical patterns of behaviour of individuals and group
- Dealing with setbacks, slowdowns, and uncertainty

Most people have some discontent regarding the current order and would like something to be altered. But those who benefit most from the current situation are those in power. They have the most influence, the most to lose, and the strongest motive to keep things pretty much the same way they are. While initiating a new system, normally some resistance is seen.

Many people in the organization, including a proportion of the management team or top people, oppose the very idea of any real change in the status quo. The reasons might be self interest, misunderstanding and lack of trust or fear of unknown, different assessment or bitter experiences of past, low tolerance for change, the power of the status quo, etc. The people somehow do not readily accept a change, and like to continue in the same situation or system, which are in practice for a long time. When the resistance to change is examined, we find that we deal two changes; a technological change and a social consequence of the technological change. Some of the methods used for overcoming a resistance are as follows.

- *Participation and involvement* – The initiators listen to the people who are involved in the change and use their advice. The participation leads to commitment, and not merely compliance of a job, however, it might take more time. Members participate both in planning and execution of change and sufficient time is provided to evaluate the merits of the change in relation to the threat to their habits, status and belief.
- *Facilitation and support* – Training is provided in new skills and sufficient time is given to employees for implementing a change. Facilitation and support are most helpful when fear and anxiety lie at the heart of resistance.
- *Negotiation and agreement* – Incentives are offered to active or potential resistors. It is particularly appropriate when it is clear that someone is going to lose out as a result of a change and his power to resist is significant.
- *Manipulation and cooperation* – Co-opting an individual by giving him a desirable role in the design or implementation of a change is a form of manipulation. This involves selective use of information and conscious structuring of events which can deal with resistance.
- *Explicit and implicit coercion* – People are forced to accept a change by explicitly or implicitly threatening them or by actually firing or transferring them. This is a risky process as people strongly resent forced change.

Four steps for bringing a change is explained in Indian *Subhaashitas*, as "*Saama*", "*Daama*", "*Bhedha*" and "*Danda*". *Saama,* the first step is to explain and convince. The second step *Daama* is to explain the consequences of not accepting a change. *Bhedha* is to differentiate or separate the people who are accepting a change from those who are not ready to change. *Danda* is to punish, the last resort for bringing a change. These are called as "*Chaturopaaya*" i.e., four techniques.

Different models are available for organizing a change. Most popular among them are Unfreeze-Change-Refreeze, change within, learning and power sharing, team management and human relation approach. Unfreezing is making people to recognize the need for a change. Once the need is felt, new method and guidelines for change are introduced and applied. Refreezing stage provides required reinforcement to ensure that new behaviour patterns are adopted permanently. The driving forces and restraining forces for a change are to be understood and balanced to have equilibrium. Internal distribution of power, internal mobilization of energy and internal communication are very important factors in achieving a change. To bring a change within, one has to concentrate on correspondence between internal and external reality, goals, values, skills and strategy. Learning helps a man to change himself. Learning takes place at all levels of life and working and not limited to any class room. Unless the seniors or leaders set examples, juniors or followers cannot learn and implement new patterns. Along with learning, there should be an enhancement of power, as successful change results from shared power and not from unilateral or delegated approaches. An approach of team management with high concern for people as well as production helps in achieving the change. This concentrates on team training, training inside the workplace for better integration between functional teams, arranging goal setting sessions, implementing plans and finally stabilizing. In the human relation approach, managers are given training to understand human problems, diagnose the need, attitude and feeling of the people, and their capabilities, and to restructure the activities to implement the change successfully.

Different approaches are adopted in managing a change. Following are some examples.

- *Information* – Providing sufficient information helps in motivating an individual to accept the change, although that itself is not a motivation factor.
- *Individual counseling and therapy* – This helps in achieving change at individual level by creating new insight, and is deeper compared to providing just information.
- *Influencing the peer group* – It is observed that individual behaviour is considerably influenced by the peers. The change process initiated in such a group is likely to be self-energizing and self-reinforcing.
- *Sensitivity training* – This training helps people to overcome the ego, understand the sentiments of others and cooperate for the common cause and to achieve the required change.
- *Feedback* – Getting feedback and analyzing helps in correcting the situation, and builds confidence among the people working. The results need to be discussed in open, so that all involved can participate.

- *Group therapy* – In this approach, it is assumed that organizational conflicts are the result of individual characteristics and the approach consists of treating group as an individual and adopting individual therapy and social psychology.
- *Systematic change* – This is highly powerful approach in changing organizations, which necessitate direct manipulation of organizational variables. The analysis is made of all the systems and changes needed are discussed and decided, and the roles and responsibilities are redefined. In this approach, attempts are made to change the entire hierarchical distribution of the decision making power in the entire organization or in the family.
- *Technological approach* – Efficiency, quality and cost are the major considerations in a technological approach. Use of new technological methods which results in major changes in the authority and responsibility relations because of rampant mechanization, automation, and on-line controls, demands multi-skill jobs and multilevel organization structure.
- *Value centered approach* – This approach considers human motivation and personal growth by changing of values and norms prevalent in the organization or society. Total change is achieved by training and educating all from top to bottom, and atmosphere is created so that members of the organization or society or family work with each other with mutual confidence and trust.
- *Structuralapproach* – This approach employs the patterns of change in authority. The changes are initiated by top and are well planned. It may be an alteration of authority by decentralization and the establishment of different profit centers or an emphasis on formal structure, controls and workflows.

There is no single simple formula for bringing a change. The guiding principles for managing a change as given by Hutton are as follows, which are key success factors.

- If you do your home work, most of the issues that need to be dealt with are already addressed in some way in your plan.
- You are not the first – many others have already done what you are attempting.
- You can draw upon the practical experience of outsiders and other organisations to help you.
- You already have access to the most important sources of information to help you figure out what to do – your people and your organisation.
- In working with individuals to win their commitment to change, start with their needs and ambitions – listen, don't preach.

- Work on building the commitment of the top key persons, and never stop reinforcing this commitment.
- Make the change process a team effort and, thus, ensure that everyone involved has the opportunity to take ownership of the process.
- Build partnerships that include all the key stakeholders; those who have the authority, the resources, and the expertise.
- In supporting the transition, set priorities and focus your efforts where they can be effective. Work with enthusiasts who will lead the way and don't waste time trying to convert those who are lost causes.
- Strive for some small tangible early successes, and make the most of these through recognition and publicity.
- Ensure that those who are affected by the changes are involved in planning their own journey.
- Strive to act as a role model for others.
- Provide information to those affected about the need for change, the means, and also the effect of change on people.
- Listen, and offer empathy for the stresses people undergo during change.
- Celebrate progress and make it fun.

By identifying the problem, analyzing the reasons, devising remedy, reinforcing the new system by carefully overcoming the resistance can bring a change in the situation, which is beneficial. It can help you to win, and you can be a winner. A supervisor, who is supposed to bring change in culture among his fellow workmen need to understand and work.

Supervisors and customer orientation

11.1 Customer orientation

Any organization if has to survive, should understand the customer needs and provide products and services to meet them. The marketing team normally has direct contact with the customers and gathers information relating to customer's expectations and needs by various means. The supervisor in shop floor is supposed to produce the materials as per customer's expectations and needs. It is therefore necessary for him to understand the requirements precisely and explain the same to the workforce in a way they can understand and act accordingly.

11.2 Customer expectations

There are number of expectations from the customer depending on the product and services being procured and the purpose of procuring. One needs to discuss with the customer, understand the purpose for which the material or service is being purchased, the way in which it is used, the culture of the people concerned, their likings and disliking, the social obligations, the legal and statutory requirements and the price customer can afford. The supervisors should discuss with the marketing personnel and get the requirements that are specific to the order. While doing this, the intended use and the quality needs considering the objectives of the product are to be understood. In number of cases, the specifications given by the customer shall not be complete, and in such cases, the technicians have to complete the specifications as per similar products and get the approval from the customer through the marketing section before starting the production. It is also customary to give a prototype sample to the customer and get feed back and finalize the specifications. In garment industries, the samples are produced at different levels and got approved by the customer before starting bulk production.

The general expectations from the customers are as follows:

❖ *Desired products* – The customer pays and buys the product required by him and hence, we need to produce the materials as per his requirements. We might have produced something fantastic, but he cannot purchase unless it is needed.

❖ *Timely delivery* – The customer needs materials for using at his end at the time when it is needed and not when not needed. It is a waste if not available in time. Non receipt of material in time can become a threat to the running of business itself; customer might have to lay off the people for no work.

❖ *No increase in prices* – The customer shall have planned his activities considering certain costs for the materials being procured. If the prices are increased in between, all his calculations shall fail, and he shall have to face losses. Hence no increase in price is accepted.

❖ *Prompt and quick service* – The customer purchases materials from us to use them. While running the materials if he finds any problem, his activity shall be affected, and hence he wants our help. We need to provide timely service.

❖ *Smooth working at his place* – The materials are purchased to work smoothly at the customer's place, and not to create problems. Hence, while deciding the process parameters and product specifications, we need to understand the purpose for which the materials are being taken and design the product accordingly.

❖ *Compensate for the losses due to quality* – Customer has purchased materials to run his business. Why he should suffer because of the quality problems in our supplies? He demands for compensation much higher than the sale value of the supplies. Hence, we should be clear on the objectionable faults or errors that can happen and design systems to prevent.

11.2.1 Analyze complaints and feedbacks

The customer complaints and feedbacks are to be taken as basis for deciding on the specific requirement of each customer. Some customers are particular about the exactness of count, whereas some are interested in smooth working at their place. Depending on the segments, some customers are more particular about the labels, whereas some are interested in the fit. Collecting data on a continuous basis and reviewing periodically can help in understanding the needs of different customers. Following table gives a matrix indicating different types of reactions for the same materials.

Customer complaint and feed back analysis matrix

Customer	Number of complaints/feed backs received from 1 Apr 2000 to 31 Mar 2009													Total	Product
	Count is fine	Count is coarse	Count CV% high	Twist is low	TPI variation	High snarling	Low RKM	High tensile CV%	Low elongation	High U%	High imperfection	Cone weight is not uniform	Low carton weight		
A	3		6		2	5				3	8	15	3	**45**	20 KHC
A	5		3		3	3	1			4	4	13	5	**41**	30 KHC
A		4	3				4		3			31	1	**46**	40 CWP
A		5	3				4		3			23	1	**39**	44 CWP
B		2	3			3					7			**15**	20 KHC
B		3			2	1					6			**12**	30 KHC
B		10	2	3				5	2	5	8			**35**	40 CWP
B		12		2		1	2		1	1	4			**23**	44 CWP
C	2		4	3	2		2				2		1	**16**	20 KHC
C	1			1		1	1				1		1	**6**	30 KHC
C	1		3				1		1		1		1	**8**	40 CWP
C	1		1				1				1		1	**5**	44 CWP
D	2			2	2		2		2		3	1		**14**	20 KHC
D	3			5			2		1		1			**12**	30 KHC
D		1		1			1					3		**6**	40 CWP
D		2		2			4		2		1			**11**	44 CWP
Total	**18**	**39**	**28**	**19**	**11**	**14**	**25**	**5**	**15**	**13**	**47**	**86**	**14**	**334**	

Above table shows that 4 customers A, B, C and D were getting the same quality yarn from the spinner XYZ. Although same quality of yarn is supplied to all, the customer A has made maximum complaints, whereas customer C has the least complaints. The customer B complained more on count being coarse, whereas for the same yarns the customer C complained the count as fine. Customer A made more complaints on cone weight variations, whereas B and C do not see it as a problem. The complaints depend on the type of machines the customers have, the product quality demanded by their customers and their management objectives. Therefore, the supervisors are required to understand the specific needs of the customers before taking the lot for production and educate the workers on the requirements.

11.2.2 Internal customer orientation

When we talk of a customer, we need to be clear; it is not only that person who purchases or receives our products and services, but also the people with whom we are regularly working, in other words internal customers. They receive our instructions, our reports, our advice, our services, various materials etc., so that they can perform their job. What we give them should be suitable to perform their work; in other words, it should meet their needs. If internal customers are satisfied satisfying the external customers is easy.

In number of cases, it is difficult to get the complete picture of the performance of our material at the customer's end. However, we can assume that our materials are performing well provided they perform well at our internal customers. The persons or the section to which we are providing products and services, i.e. our next man in the organization, is termed as an internal customer. Internal customers need good quality and services so that they can produce good quality products to give to external customers. We should ensure that our internal customers are satisfied with the quality, delivery and services. External customer cannot be satisfied unless internal customer is satisfied.

11.2.3 Internal customer interface

Internal customer interface is a mechanism of understanding the needs of internal customers and to arrive at a consensus. The senior members from the internal suppliers and internal customers sit and discuss the quality requirements. The meetings shall be periodic and normally at fixed time and venue.

Methodology

- The customers and suppliers sit across a table.
- The customers shall spell out their problems and the requirements from the products and services provided to them.
- The suppliers might not be in a position to fulfill all the demands made by customers.
- The suppliers shall note down all the points. They will not give immediate reply
- The suppliers shall sit together and discuss among themselves as to what are the things they can commit, and how much time is required for them.
- They also discuss as to what cooperation from customer is required to fulfill their demands

- Once all the members in the supplier's team come to a conclusion, they will come back to customers
- The suppliers shall group the demands as:
    - → Immediately possible and shall be attended
    - → Requires certain time and shall be attended
    - → Customer need to cooperate to make this happen
    - → Not possible
- The customer team shall discuss among themselves regarding the time, cooperation and alternatives for the demands not accepted
- After discussions, the customers and suppliers shall come to a consensus.
- Agreement shall be made, and documented in procedures and work instructions.
- Follow up shall be made by internal quality auditors for the fulfillment of the agreement.

11.3   Understanding customer perception

The textile and garment industry have own specific features. The industry has various sections; some catering their products and services to other manufacturers for adding value, whereas some catering directly to end customers. Where, one supply his products to another manufacturer, the customer satisfaction shall be depending on the precession needed in the products and their properties, the timely deliveries, the timely support by the technical team and the price. Where the materials are being used for industrial applications and for specified functions, the customer requirements shall be more or less clear to the supplier, whereas when catered directly to the final customer like garments, it is very difficult to understand the real needs of each and every individual customer.

There are different types of textile units supplying intermittent material for further conversion or value addition at a different place. Following are some examples.

- ➤ Sliver making and suppliers of inter bobbins for decentralized or hand spinners.
- ➤ Spun single yarns suppliers for weaving, knitting or doubling factories.
- ➤ Doubled and cable yarns suppliers for weaving, tyre cords, technical textiles and industrial applications.
- ➤ Woven grey fabrics suppliers for various applications.
- ➤ Knitted fabrics suppliers for garment units.
- ➤ Garment washers doing job works.
- ➤ Fabric printing units.
- ➤ Yarn and fabric processing units.

The products produced in the above units need to be processed or converted further to make it suitable for use by the final customer. The processed fabrics either woven or knitted can be sold directly to the end users so that they can get the garments stitched as per their will. These fabrics can also be sold to garment factories for producing readymade garments.

Where the materials are procured for further conversion activities, the main requirements will be the smooth trouble free working, on time delivery, affordable price and the specified quality. The product parameters shall be directly measurable with internationally accepted test methods. The customer can clearly specify the reasons for his dissatisfaction and the supplier is in a position to correct it. In case of materials sold in retail market for the end user, it is very difficult to understand the real reason. Following are some examples:

> A lady rejected a particular saree as her neighbour was having a similar one.
> A shirt was rejected because it looked more traditional.
> A good dress was rejected by a girl as she already had one set similar to that.
> A particular style of jeans could not be sold in the market because a movie hero changed his style.
> A boy rejected a suit because his girl friend had made a comment on that type of suit worn by somebody.

There are number of such reasons, where a technician is in confusion and cannot take an action for improving the customer satisfaction. In case of materials going for specific end use, a technician can work and improve the consistency in the supply. The technicians should understand the impact of various technical parameters of the product on the performance at customer's end.

11.4   Communicating customer needs down the line

Communicating the customer needs down the line is very important to ensure adhering to the quality needs. The normal practices include circulating the techpack or the agreed specifications to all production in-charges and supervisors, conducting order clarity meeting involving marketing, production planning and control, production personnel, maintenance personnel, quality control and industrial engineering personnel. The techpack or the specification sheets give the stated needs of customer, whereas, in an order clarity meeting, one can discuss and understand the intentions and unspecified needs of the customers also.

Some of the companies display the customer name and previous complaints on each machine so that the workers can be alert and prevent the repetition of a complaint. The quality control in-charge shall be monitoring with the marketing personnel and list out all the feedbacks and complaints and give to production in-charge before starting a particular lot for a specific customer. Some companies even display the complaint samples sent by the customer.

When orders are received from a new customer, the marketing people address the production and quality control persons regarding the exact need of customers. In garment industries conducting order clarity meeting has become a ritual where the merchandiser, quality control person, industrial engineer, production person, maintenance person, production planning person, and the sourcing persons sit together and understand the finer details of the customer requirements. There are instances of a representative from customer taking part in the order clarity meetings.

Quality management and assurance

12.1 Expectations from supervisory staff

The top management stands guarantee for the quality of products and services to their customers and expects the supervisory staff to manage the production with required quality and deliver them to customer while maintaining the costs at affordable levels. The supervisory staffs are to be always on toe, plan and monitor the activities and ensure that the products and services are delivered as needed. The supervisory staff should be thorough with the concepts of quality management and quality assurance. The quality assurance is a part of quality management. The quality management has four wings viz. quality planning, quality control, quality assurance and quality improvement.

12.2 Quality planning

The important factors contributing for the success of a manufacturing process in an industry are proper quality planning, product design, process design, process monitoring, logistics and services.

Quality planning is the process of planning the activities in order to achieve the goals of meeting the customer requirements in time, with in the available resources. Understanding the customer's needs is the first step in the process. It is defined in ISO 9000:2000 as "Part of quality management focused on setting quality objectives and specifying necessary operational processes and related resources to fulfill the quality objectives". In textiles, the ultimate consumers and also the men involved in retailing are normally not technicians, and hence the requirements are not clearly explained as required to a shop floor technician. Although in some cases, technicians are employed for identifying the specific needs, the interpretation changes, and the production personnel get a different message. The over-specification is a common phenomenon so as to overcome failures, which is resulting in increased expenses.

Let us take an example of 20s carded hosiery for knitting purpose. The customers ask for the best yarn, and often refer to a benchmark like uster Statistics and demand for 5% level or 25% level. They never try to realize, whether that quality is required for the product manufactured and the technology adopted. One should understand that, uster 25% indicates the quality level achieved by the top 25% of the mills who participated in the survey. Normally the mills with new equipments participate in such surveys and others hesitate. All the mills on earth are not participating. Further, unless Ms. Zellweger uster requests a mill to participate, the mills will not participate. That being the case, we should try to know, what is happening to the yarns made by other mills, which are inferior to uster 25% level. If that yarn can be used to get the required end product, why we should not use it? Whether the quality of yarn required for all end uses are same, for e.g. hand knitting, slow speed knitting, high speed knitting, etc, or for the different products with single jersey, pique, interlock, rib etc, for t-shirts, socks, hand gloves, sanitary napkin covers, under garments, rexene coated covers etc? It should also be noted that uster gives statistics for each parameter separately, but the customers interpret that a single particular yarn had all the parameters at uster 5% or 25% level. A mill with old machines, which cannot run at a high speed gets a higher elongation compared to modern mills with high speed working and hence, ranks in better level like uster 5%. So we should think before coming to a conclusion. The shop floor technician should be clear on what quality parameters could be committed and explain the same to the marketing department and the customers before starting the production activities.

Once the identification of a need is clear, one can plan the raw materials, processes needed, machines to work, process controls needed, inspection and testing, training needed for operators, starting and ending date of production, infrastructures needed, information required, etc. The technician should be clear about the technical capabilities of each machine and the skill level of each operator so that he can plan the machine and men allocation to achieve the targeted production and quality.

The quality objectives of each product are to be determined, which shall help in deciding the specifications. For example, 20s carded hosiery yarn required for producing single Jersey on a high speed knitting machine should have the quality objectives, something like the following.

- Sufficiently even to get a uniform appearance of the fabric.
- Sufficient and uniform strength to work at high speeds on knitting machine without excessive breaks.
- Minimum lint generation while knitting at high speeds.
- Free form dead cottons and immature fibres to get a uniform depth of dyeing.
- Flexible and smooth to move freely in the needles.

- Uniformity with a low twist to have smooth feel and without spirality.
- Uniform density of cones to have uniform tension and even knitting.
- Equal length of yarn on cones to have lower wastages.
- Count as required to get the GSM as specified by the ultimate customer.

By going more in detail, one can work out the quality objectives more precisely, and fix the parameters, which are measurable. The quality objectives of a product must address the following:

- For what end use the customer is asking this material?
- What are the critical quality requirements of the end product?
- What is the intended application of the product with ultimate consumer?
- On what machines our product is likely to work, i.e. high-speed machines, super high speed machines, slow speed machines, automatic machines, etc.
- What is the work environment at customer's place; like humidification, temperature, dust level, skill of operatives, the management culture, etc?
- What are the applicable regulatory requirements for the product from the point of view of product safety, user safety, trade regulations etc?

To ensure that parameters are met all the time, internal targets are to be fixed in each process by carefully assessing the process capability of the plant in operation. The supervisory staff in the mill has to work out internal targets for each type of machine they have. The process control norms also depend on the process and machine capability. Once the targets are fixed, the process and procedures can be established, and resources can be planned and provided. The inspection and test plan depends on the customer requirements and not on the process capability or machine capability, as the customer is least interested on the problems we have, but is concerned about his getting correct material in time. If we give something better than the expectation of customer with out increasing the price, he shall be happy, but shall not accept any deviation in quality in future supplies. Hence, while deciding the internal norms and acceptance criteria, one should work very carefully. The supervisors should be careful and avoid over designing of the product and also removal of excess wastes to achieve the quality. The planning for production should be done in such a way that customer gets what he expected, and we do not over do any thing. Therefore, the important steps in quality planning are:

- Understanding customer needs and deciding quality objectives of

the product.
- Identification of processes needed.
- Identification of machines to work.
- Determination of process controls and norms.
- Determination of inspection and testing and acceptance criteria.
- Planning the starting and ending time.
- Identification of persons to operate and control.
- Identification of training needs.
- Working out infrastructure requirement.

The plans need to be documented for effective follow up while manufacturing and for monitoring the process. An effective quality plan helps in achieving the results as expected in product and services. It is better to prepare checklists for follow up, so as to ensure that nothing is left out from what was planned.

12.3   Controls and checks

In order to achieve the required performance, we need to identify the areas, which are to be controlled, and what are to be checked and ensure proper control. The control points if controlled should lead to the achievement of the result in the key result area, and finally the company objectives and goals. The checkpoints are process related, where as the control points are result related. It is necessary for the concerned heads of department to analyze their processes very carefully and decide on the area where they need control. Once the area to be controlled is clear, one can identify the points to be checked to verify whether the process is in control. The quality control personnel concentrate on whether the product and process are as per the agreed norms by verifying the check points suggested, whereas the production and maintenance personnel concentrate on control points and ensure that they are all in control.

Control points	Check points
Process parameters	o Level of adherence to process parameters. o Calibration of equipments monitoring the process. o Suitability of process parameter decided to get the results. o Product quality achieved
Selection of raw materials	o Quality of materials received. o Handling and storage systems. o Combination of materials taken for production
Selection and training of employees	o Competency levels of men available and employed. o Process performance. o Work practices. o House keeping practices.

Maintenance of machines	o Adherence to maintenance schedules.
	o Suitability of maintenance schedules and plans for the production and quality expectation.
	o Condition of machine parts.
	o Maintenance practices.
	o Results of maintenance.
Rejection rates	o Clarity of acceptance criteria among all on the shop floor.
	o Whether process parameters are as per standards specified?
	o Actual Rejections; machine-wise, shift-wise, operator-wise and material-wise.
Delivery schedule	o Whether production started in time?
	o Utilisation of machines.
	o Productivity of each machine.
	o Whether quality approved?
Inventories	o Material in process.
	o Stock value of materials in stock at Raw material stage, finished goods and material in process.

5-S concepts can be used in deciding the control points and checkpoints in process monitoring. This helps in identifying what is required to be maintained and kept, and what is to be discarded, which becomes the basis for the design of work place layout and flow of process.

After deciding the checkpoints and control points, allocate the responsibilities for the people on the shop floor to check the process and take suitable corrective and preventive actions to bring the process in control, and to maintain it. These points need periodic review, and modifications as the systems improve. For example, in a scutcher at blow room we needed to control the linear density of laps. Therefore, a uniform length for all laps was fixed and weighed physically. When we change the system to chute-feed, there is no question of physically weighing the laps, but we need to ensure the required pressure and flow of materials. Therefore, we need to monitor the pressure switches. Similarly, we need to think of all the changes we are doing, may be as a part of technology up-gradation or as a preventive action for certain potential problems, considering the market feed back or by the trend analysis of the company performance. Extent of control depends on the expectation of customers, and hence, it cannot be same for all mills. Depending on the market feed back, one needs to continuously upgrade the control points and check points for his processes.

12.4 Quality assurance

Quality assurance is defined as a planned and systematic pattern of all actions necessary to provide adequate confidence that the product optimally fulfils customers' expectations, meaning that it is problem-free and well able to perform the task it was designed for. Quality assurance concentrates on identifying various processes, their interactions and sequence, defining the objectives of each process, identifying the key result areas and measures to measure the results, establishing the procedures for getting the required results, documenting the procedures to enable everyone to follow the same, educating the people to implement the procedures, preparing standard operating instructions to guide the people on work spot, monitoring and measuring the performance, taking suitable actions on deviations and continuously improving the systems.

The main elements of quality assurance are as follows:

- Well defined procedures to achieve the objectives
- Documentations to prevent any lapses in implementing procedures
- Adequate records for taking decisions and actions
- Adequately competent and trained employees
- Periodic reviews to ensure that the system is process and adhered

The role of technical supervisor is very important in assuring quality to the customers by adhering to the systems and monitoring the same round the clock. A good technician never allows a deviation in procedure unless it is approved by the top management under the circumstances and responsibility is taken by the top.

12.5　Quality improvement

Improvement is a continuous process and cannot be stagnant. Quality improvement is systematic approach to reduction or elimination of wastes, work-back flow, reworks and losses in production process. Taking timely corrective actions and preventive actions to the deviations found, the feedbacks and complaints received are the starting point for any improvement approach. The role of technical team is very important in any organization.

The improvements may be by eliminating the source of errors or by stretching the targets for achievement after each achievement. Keeping a continuous watch on the performance and acting in time is very important for a supervisor. Analyze the reasons not only for failures but also for success. Whenever you achieve a result better than your anticipation, make analysis to find out the factors responsible for the success and try to develop systems to sustain it.

Quality improvement is a never ending process. The customer's needs and expectations are continuously changing depending on the changes in technology, economy, political situation, ambitions and dreams, competition, etc. One needs to go on analyzing the facts and identify potential problems in advance and take necessary precautions to survive in the competitive world. We need to consolidate all steps we take for improving a situation and implement them uniformly organization-wide to get consistent improvement. We need to be always on the toe for facing challenges, benchmark the best and try to re-engineer the activities and the products to beat the competition. The use of Five Golden Questions for self assessment can help in continuous improvement, which is as follows.

1. Are we having a procedure?
2. How do we ensure it as the best?
3. How did we implement?
4. Did we achieve the results as anticipated?
5. How do we compare with our competitors?

12.6   Visual management

It is essential to have discipline while we wish to sustain the performance. Without a discipline, the system shall collapse. The visual management principles help in identifying and removing the unwanted materials and process and providing space for the required activities. The concepts of visual management are derived from the 5S concepts which are as follows.

1. *Seiri* → *Sorting out* – Find out which of the materials and sub process is required to achieve the results, and which are not contributing. Eliminate non-contributing materials and processes. We are having lot of materials and doing a lot of works that are not contributing either to improve the quality or to reduce the cost, but are adding the costs. Why should we keep them? The concept of Seiri is 'remove the scrap away and you get space released for others'. Move items useless to you but likely useful to others to a common place which shall help in eliminating the unwanted purchases. Store away the materials used less frequently and retain the materials frequently used near the work place. This will avoid confusions and mix ups.

2. *Seiton* → *Systematic arrangement* – Arrange the materials and the processes in such a sequence that the operation costs shall be least with the assured results. This includes the re-layout of processes and bypass systems and also organisation structure. Systematic arrangement includes the fixation of place for each item, colour codification and signboards for identification and display of standard work practices and instructions. It reduces the searching time and stress, and the excess inventory shall be visible which helps in controlling and finally reduces the cost of inventory.

3. *Seiso* → *Spic and span cleanliness inspection* – Keep the work area always clean so that any deviation or mistake can be seen. Design the process control systems to get the optimum efficiency with assured quality. The inspection and testing suggested should be able to help the people to take corrective action. There is no meaning of getting test results, without knowing how to control. This concept includes the cleaning of everything at work place and inspection for abnormalities during cleaning and routine maintenance, thus reducing the chances of breakdowns.

4. *Seiketsu* → *standardizing* – Standardize the layout, place for each material and processes to get consistency in quality and productivity. Remember that success lies in consistency and not in getting some record results. This includes the visual display of standard operating systems maintained and monitored by regular audits, and improved. This helps in achieving the targets.

5. *Shitsuke* → *self-discipline and training* – Monitor the process on a regular basis, and provide education and training to the people involved in the operations so as to ensure uniform working. In the absence of suitable discipline, any system shall fail. This emphasizes on the inculcation of self-discipline by top management, and imparting of tactical and strategic training. This shall improve the morale among the employees.

12.7   Six sigma and zero defect concepts

The success of a company depends on the consistency in the quality, and not a very good quality at some time. The customers are more interested in their performance and need consistency in what they procure so that they can set their process and run without interruptions. The technicians have to work for reducing the variations and defects.

The quality control is the process of checking the process and products with an intention of preventing nonconforming materials from going to the customer. But it is not possible to check the quality of each and every material produced. These checks can only prevent a bad material from going out but cannot prevent it from being produced. The reasons for poor quality or variations in quality levels can be attributed mainly to improper or variations in raw materials used, the faulty work methods adopted, improper maintenance leading to poor machinery conditions, improper selection of process parameters, appropriate technology not being adopted for the process and the product requirements, and not following the accepted systems religiously.

The six sigma concept is on reducing the variability in the process that can lead to reduction of variability in the product. Some variations are termed as normal variation. Even though there is no change in raw materials, process, people, machine, setting, etc, there shall be some variations in the products. This depends on the capability of machines, men and process. When the frequency is plotted we get a distribution on both the sides of mean in a uniform way. Normally 99% of the readings fall with in ± 3 sigma limits, where sigma is the standard deviations. In the Six Sigma concepts, efforts are made to reduce the value of sigma in such a way that the acceptable tolerance of the customer is with in ± six sigma of our process. It means the standard deviation is reduced by over 50% than the present level. This is done by improving the process capability not only by improving the technological aspects, but also the work culture.

The strategy for implementing six sigma concepts includes define, measure, analyse, improve and control. Improvement projects are integrated with the goals of the organization and a "divide and conquer" approach is used as opposed to continuous process improvement. Important steps in implementation are:

- Measuring and understanding present level of performance and quality.
- Identifying the areas having variations.
- Analysing the reasons for variations.
- Working out systems to reduce variations.
- Educating and training the concerned in the improved systems.
- Developing self monitoring teams at all places.

The central idea behind six sigma is that if you can measure how many "defects" you have in a process, you can systematically figure out how to eliminate them and get as close to "zero defects" as possible. Unless you measure you cannot improve.

12.8   Lean management and waste reduction

Lean management concepts try to reduce the non-value added activities in a process and also to balance the process flow to have lowest possible delays and inventory. It focuses on maximizing process velocity and provides tools for analyzing process flow and delay times at each activity in a process. It concentrates on the separation of "value-added" from "non-value-added" work with tools to eliminate the root causes of non-valued activities and their cost and attacks the 8 types of waste or non-value added work The 8 types of wastes are:

- *Wasted human talent* – Underutilization of a man's capability makes him dull. It is damage to people. Human motions involve time, and unnecessary movement is a waste. It could have been utilized for useful activities. Human potential is the biggest asset of any organization. A bad car can be driven safe by a good driver but not a good car by a bad driver. The melody of music depends on musician and not on the instrument alone. Idles brain is devil's workshop; do not allow the devil to grow.

- *Defects* – Stuff that is not right and needs fixing. Customers have paid money for defect-less product and not for products with defects. Hence, product with defects is a waste.
- *Inventory* – Any material waiting to be worked is a waste. It is adding value when being worked and not when it is idle. Inventory may be the raw materials and accessories waiting in stores or godowns, the materials in process and the finished goods in warehouses.
- *Overproduction* – Producing more than required or the producing before time add to the idle inventory of finished goods after spending for conversion. Over production results in unwanted stocks that cannot be sold, leading to recession. It reduces the demand. Material stock adds to the costs of carrying, but its value goes on decreasing as the time passes. It is better not to produce rather than producing when we are not able to sell.
- *Waiting time* – People waiting for anything is a waste. People cannot produce while waiting, but the company has to bear their expenses. This problem is more when the productions are not balanced in a line or there are activities of outsourcing. Just in time approach tries to reduce the material waiting time by ensuring that only the required material is supplied in time by proper planning and balancing of the material flow and continuous interaction within the supply chain.
- *Motion* – Unnecessary motion is a waste. The motion should be to add value and not to add only cost.
- *Transportation* – Moving people and stuff from one place to another does not add to the value, but adds to the cost. It cannot improve quality but can damage. Transportation should be avoided as much as possible
- *Processing waste* –Stuff we have to do that doesn't add value to the product or service we are supposed to be producing. For example, the extra projections given to handle the materials, the extra length wound to facilitate threading or taking the materials inside the machines, the projections or the sides we trim and so on.

12.8.1 Wastes in textile and garment industry

In textile and garment industry, the wastes can be grouped as product wastes and process wastes. Product wastes are the defective materials produced, excess produced and surplus fed than actual requirement to get uniform product. The process wastes are inevitable wastes during process, extra projections for handling, and the materials like water, colours, chemicals, coolant, and lubricants etc., used for the process. The supervisors should work to reduce both types of wastes. The defective production can be reduced by understanding the product, designing the proper product, selecting the correct materials, establishing appropriate procedures and parameters, training the people to produce the required quality materials and monitoring the process to avoid defects. The product wastes are very costly; they would have got same price as fresh material if it were not defective. Process wastes can be reduced by proper selection of material and designing the process. For example, when pattern and marker is designed well, the marker efficiency shall be high and fabric cut wastes shall be less.

Any reprocess is additional cost, which restricts capacity and causes delay in deliveries. The major loss is losing the confidence of customers. If there is no reprocess, that much additional materials would have been produced adding to the profit. One should note that producing correct quality in the first attempt is easy rather than in reprocessing and chances of failure is always more in reworking. It is true for spinning, winding, wet processing and garment making, where as normally there are no chances of reprocessing in weaving and knitting. There are chances of redrawing the beam in case of improper fabric parameter or poor working in case of weaving, which is a very costly affair.

The wastes can either be visible or invisible. The visible wastes are those where the waste materials can be seen, collected, weighed or measured and accounted. In case of invisible wastes, the wastes escape into atmosphere, cannot be seen or collected. Accounting is done by indirect way of accounting the good materials and comparing with the consumption. One has to work to reduce both types of wastes.

The wastes also can be grouped as saleable wastes and useable wastes. Useable wastes are usable either in the same product being manufactured or in any other product in the same company. The wastes that are not useable in the same company are either sold to others or scrapped/destroyed and are termed as saleable wastes. Sometimes, people try to takeout more saleable wastes in order to achieve certain quality level, but it adds to the cost of material being produced. One has to work for reducing both the groups of wastes.

Downtime is unplanned stoppage due to breakdowns, short of materials, short of men, failure in power supply. Down time reduces efficiency, increases cost, delays the delivery and also can spoil the quality, especially in special processes. The supervisors should work to reduce the downtime as minimum as possible with a target of zero downtime.

12.8.2 Waste disposal

What ever may be the waste, it is a waste. Please avoid generation of waste. If waste is generated, learn to dispose them without harming others.

- Segregate the wastes and group them as per their properties. If the wastes are mixed up, the rate fetched shall be same as that for the lowest value component. For example, if good cottons are mixed up with sweeping waste, it shall be valued as sweeping waste and not as cotton.
- Have designated collection points depending on the nature of the wastes. This shall prevent mix up of wastes, and also shall help in easy disposal.
- Have suitable handling and packing process to avoid liberation of micro dusts while handling loose cotton wastes. At present pneumatic systems are available to collect wastes from the source of waste generation, filter them and pack them in bales without the intervention of any human.
- Identify the wastes that are reusable and collect them separately. Send the reusable wastes to the concerned users with proper accounting. For example, comber noil taken out as a saleable waste can be used in the same mills as a component in coarse open end yarn. The end bits after cutting a role in garment factory can be used for re-cutting and replacing the defective panels if any.
- Locate the authorised agents for disposing harmful wastes. The wastes should not create problem of pollution or health hazard for the public. The governments have authorized certain agents to handle the hazardous wastes. The wastes are to be allowed to collect by such agents only.
- While discharging waste water, effluents etc, care should be taken to treat them and dispose to the identified location, which is in line with the direction of pollution control board.
- While disposing the empty containers of chemicals, ensure that they are washed thoroughly.

Practical competency and underpinning knowledge

13.1 Defining competency and knowledge

For doing any job, the person should have the required competency and certain knowledge about the job. Although the designations for the staff and workers are similar between mills or factories, the technology and the product range are different. Hence, the competency and knowledge specified by one company need not suit another company. These are to be defined by the each company depending on their product range, manufacturing facility, the market requirement and the vision of the company. There are different types and levels of supervisory activities in a textile or garment industry. There are certain factors that are common for all supervisors and can be referred as general points. The general competencies are as follows:

- Taking round of the work area before the start of the shift and observing the working.
- Taking charge from the previous shift supervisor with details of work in progress and works to be done further.
- Understanding the production plan and allocating the machines for different activities.
- Understanding the machines allotted for various mixings and products and deciding on the changes to be done if needed.
- Checking the quality of materials being produced by taking rounds and informing the concerned maintenance personnel for correcting in case of poor quality.
- Allocating the workers on the machines considering their skills and workloads agreed.
- Checking the productions periodically and taking action where it is low.
- Working out the changes to be made for product mix and giving instructions.
- Monitoring the humidity and temperature as per requirement by

coordinating with the concerned engineering operator.

- Checking the conditions of accessories and taking only the suitable ones.
- Checking the colour codification and ensuring there is no confusion or chances of mix-ups.
- Checking the sweeping wastes for preventing good materials going in the wastes.
- Counseling a habitually absenting worker and referring to HRD if needed.
- Counseling and influencing a low producing worker to produce as per norms.
- Recording the stoppages and working out the production loss due to stoppages.
- Recording the activities in the log book (Report Book) before handing over charge to next shift supervisor.
- Ensuring use of safety gadgets like caps, masks, gloves and shoes by all concerned and also by self as per the work requirement.
- Verifying the safety stop motions provided and getting them attended in time.
- Maintaining records of production, summarizing and submitting to management.
- Participating actively in mock drill for fire fighting and first aid and get prepared to handle any such events.

The generally required knowledge is as follows:

- Importance and functions of various machines, mechanisms and infrastructure in the section.
- Production balancing – Importance and methodology for different product combinations.
- Role of humidity and temperature in maintaining quality and productivity.
- Work loads, work allocation and standard working conditions appropriate to the section.
- Calculation of production and efficiency, the industry norms. Factors affecting productivity.
- Colour codification and its importance.
- Roles and responsibilities of a supervisor.
- Basic supervisory skills – Listening and observing, communicating, counseling, taking charge, reporting and motivating.
- General management knowledge of managing subordinates, coordinating with workshop, electrical department, stores and production.

- Standing orders and discipline in working.
- Precautions to be taken while working.
- Importance of cleanliness and personal safety.
- Fire fighting and first aid.
- Safety gadgets used in the factory and the workplace.

13.2   Requirements for different supervisors

Following are some examples, which can be used as a guide while providing training to the supervisory staff. For first 2 examples in production and maintenance, all competencies and knowledge are written, and for the remaining, only specific requirements are written. The general requirements are to be added for all while conducting training and preparing course curricula.

13.2.1 Supervisor – production spinning preparatory

a) *Practical competencies*

- Taking round of the work area before the start of the shift and observing the working. Noting down the machines with short of materials, poor working, quality and housekeeping problems.
- Taking charge from the previous shift supervisor, noting the mixings being worked, the machines allotted, the hank organization, the production completed so far and the production needed further and special instructions if any.
- Understanding the production plan and allocating the machines for different activities. Understanding the balance production needed in case of running out materials.
- Understanding the machines allotted for various mixings and deciding on the changes to be done if needed. Verifying the stock of material in process and deciding on the machines to be worked.
- Verifying the stock of various mixings and working out the mixing requirement for the shift and arranging for the same.
- Checking the quality of materials being produced in each process by taking rounds and informing the concerned maintenance personnel for correcting in case of poor quality.
- Allocating the workers on the machines considering their skills and workloads agreed.
- Checking the productions periodically in all machines and processes like blow room, carding, draw frame, combers and speed frames and monitoring the material flow and process balancing.
- Working out the changes to be made for change in hank or product

mix and giving instructions; like the wheels to be changes, pulleys and speeds, setting, settings, selecting condensers and spacers, and checking the wrapping after making a change.

- Working out the expected productions depending on the speed and hanks and explaining to workers regarding the productions to be achieved in the shift.
- Monitoring the humidity and temperature as per requirement by coordinating with the concerned engineering operator.
- Checking the conditions of all material containers like lap rods, cans and springs, spools for combers and empty bobbins for speed frames and removing the bad ones.
- Checking the colour codification and ensuring that there is no confusion or chances of mix-up.
- Checking the sweeping wastes of different sweepers before the wastes are put in bags for preventing good cottons going in the wastes.
- Checking the contaminations collected in cotton and tracing the bale number and lot number and informing the cotton purchase section through the assistant spinning master.
- Counseling habitually absenting workers and referring to seniors and HRD as needed.
- Counseling and influencing a low producing worker to produce as per norms and monitoring the production till he reaches the required level.
- Addressing any disputes relating to working between employees or sections.
- Recording the stoppages and working out the production loss due to different types of stoppages.
- Recording the activities in log book (report book) giving details of mixing-wise machines worked, mixings made and used, shortage or excess of materials, the working levels, problems faced and special instructions to be followed.
- Ensuring use of safety gadgets like caps, masks, gloves and shoes by all in the section as needed for the operation.
- Verifying the safety stop motions in all the machines and facilities provided and getting them attended.
- Maintaining records of production, analysing reasons for low production, making a summary and submitting to management.
- Participating actively in mock drill for fire fighting and first aid.

b) *Under pinning knowledge*

- Importance and functions of various machines and mechanisms used in spinning preparatory machines.

- Production balancing – importance and methodology, allowances for process wastes at different processes.
- Role of humidity and temperature in maintaining quality and productivity. Basic knowledge about operating and maintaining a humidification plant.
- Workloads, work allocation and standard working conditions.
- Calculation of production and efficiency, the industry norms. Factors affecting productivity.
- Colour codification and its importance. Precautions to be taken while allocating colour codes to a new product.
- Roles and responsibilities of a supervisor.
- Basic supervisory skills – Listening and Observing, Communicating, Counseling, Taking charge, Reporting and Motivating.
- General management knowledge of managing subordinates, coordinating with workshop, electrical department, stores and production
- Standing orders and discipline in working.
- Precautions to be taken while working.
- Importance of cleanliness and personal safety.
- Fire fighting and first aid.
- Safety gadgets in a cotton factory.

13.2.2 Supervisor – ring spinning

a) *Practical competencies*

- Taking round of the ring frames and winding area before the start of the shift and observing the working and stocks.
- Taking charge from the previous shift supervisor regarding the machines working on different counts, the doff balance at winding, supply of back process materials, and specific instructions if any.
- Understanding the production plan and allocating the machines for different activities.
- Understanding the machines allotted for various counts and types and deciding on the changes to be done if needed.
- Verifying the stock of various Inter bobbins and advising the preparatory supervisor.
- Verifying the stock of Ring frame doffs in winding and assessing the availability of empty bobbins for spinning and advising the winding supervisor.
- Working out spinning production plan to maintain minimum stock of doffs and ensure supply of empty bobbins.
- Checking the quality of materials being produced by taking rounds

and informing the concerned maintenance personnel for correcting in case of poor quality

- Allocating the workers on the machines considering their skills and workloads agreed and the counts working on the machines.
- Checking the productions periodically on all ring frames.
- Checking the piecing quality of individual siders and doffers from time to time.
- Checking the cleanliness at drafting zones, spinning creel and other critical places for getting quality.
- Working out the changes to be made for change in count or product mix and giving instructions like the speed, traveller, spacers, wheels to be put and so on.
- Planning the materials needed in advance for a count change and ensuring minimum time for count change.
- Monitoring the humidity and temperature as per requirement by coordinating with the concerned engineering operator.
- Checking the conditions of doff baskets/crates, empty bobbins and segregating the bad material.
- Checking the colour codification.
- Counseling habitually absenting workers and referring to HRD if needed.
- Counseling and influencing low producing workers to produce as per norms.
- Recording the stoppages and working out the production loss due to various stoppages and taking corrective actions to reduce stoppages.
- Recording the activities in the Log Book (Report Book) regarding the machines worked, changes made, quality issues, doff stocks at winding and special instructions if any.
- Ensuring the using of safety gadgets like caps, masks, gloves and shoes by the concerned.
- Verifying the safety stop motions and getting them attended.
- Maintaining records of production, analysing the reasons, making summary and submitting to management.
- Arranging and participating in mock drill for fire fighting and first aid.

b) *Under pinning knowledge*

- Importance and functions of various mechanisms used in Ring spinning section.
- Production balancing – Importance and methodology; allocating ring frames to meet the winding requirements.
- Role of humidity and temperature in maintaining quality and

productivity. Basic knowledge of operating a humidification plant to get required humidity and temperature.

- Workloads, work allocation and standard working conditions in ring frames like number of sides or spindles per sider, number of spindle doffs per doffer etc.
- Calculation of production and efficiency, the industry norms. Factors affecting productivity in ring frames.
- Colour codification and its importance. Precautions to be taken while allocating colour codes for a new count.
- Roles and responsibilities of a supervisor.
- Basic supervisory skills – Listening and observing, communicating, counseling, taking charge, reporting and motivating.
- General management knowledge of managing subordinates, coordinating with workshop, electrical department, stores and production.
- Standing orders and discipline in working.
- Precautions to be taken while working different fibres and counts.
- Importance of cleanliness and personal safety.
- Fire fighting and first aid.
- Safety gadgets in a cotton factory.

13.2.3 Supervisor – open-end spinning (rotor spinning)

a) *Practical competencies*

- Understanding the production plan and allocating the machines for different activities by taking stock of materials in packing and further processes.
- Understanding the machines allotted for various counts and types and deciding on the changes to be done if needed.
- Verifying the stock of various draw frame cans and advising the preparatory supervisor.
- Working out spinning production plan and arranging for required empty tubes.
- Checking the productions periodically on all Open end machines and taking action on low producing machine.
- Checking the piecing quality of individual siders from time to time and ensuring that rotors are cleaned before piecing where automatic cleaning and piecing are not available.
- Checking the cheese quality produced on each rotor and getting the cheese holders adjusted.
- Planning the materials needed in advance for a count change and ensuring minimum time for count change.

- Checking the conditions of cheese trolleys, empty tubes and segregating the bad material.
- Checking the colour codification.

b) *Under pinning knowledge*

- Importance and functions of various mechanisms used in Open-end spinning machines.
- Production balancing – Importance and methodology.
- Factors affecting productivity in rotor spinning.
- Factors affecting quality in rotor spinning.
- Working out production, norms for production.

13.2.4 Supervisor – maintenance – spinning preparatory

a) *Practical competencies*

- Taking round of the work area before the start of the shift and observing the working. Taking charge from the previous shift supervisor.
  - Noting down the machines stopped for repairs and the type of problem.
  - Understanding the quality complaints in the machines.
  - Understanding the works done till now and the works pending in the machines stopped for repairs or for maintenance works or modifications.
- Understanding the production plan and preparing maintenance plan and allocating people for different activities.
- Understanding the machines allotted for various mixings and deciding on the parameters to be checked while doing maintenance like condition of beaters, grid bar settings, the wire points to be mounted on cards, quality of wire points, card settings, half laps to be put on combers, comber settings, settings at draw frames, functioning of stop motions, top arm settings in speed frames, the buffing quality of the top rollers in draw frames, combers and speed frames and so on.
- Verifying the stock of various spares, accessories and lubricants and working out the indenting plan and placing indents.
- Referring the machinery catalogues and identifying the correct spares needed.
- Checking the quality of materials received at stores, for e.g. bearings, wheels, arbours, machine spares, card wires, belts, brushes, spanners and other tools, etc.

- Allocating the workers for different tasks considering their knowledge, skills, maturity and workloads agreed.
- Checking the maintenance activities in all preparatory machines like blow room, carding, draw frame, combers, speed frames and the ancillaries like trolleys, filters, and roller covering and so on.
- Referring to process parameters and working out the changes to be made for change in hank or product mix and getting the changes made by the concerned maintenance workers.
- Checking the conditions of machine parts while they are being cleaned/scoured or overhauled and getting the worn-out parts replaced.
- Counseling habitually absenting workers and referring to HRD if needed.
- Counseling and influencing a poor performing worker to produce as per norms.
- Monitoring the stoppages due to breakdowns and analysing the reasons for breakdowns and taking precautionary measures.
- Monitoring the mounting activities in cards, comber half laps and various beaters.
- Monitoring the cot mounting and buffing activities
- Conducting the tool audits i.e. the tools used for maintenance like spanners, top arm gauge, lubricating and flushing pumps, buffing machines, mounting machines, etc.
- Recording the activities in the log book (Report Book) and updating the Machine History book
- Ensuring the use of safety gadgets like caps, masks, gloves and shoes by all maintenance workers.
- Verifying the safety stop motions and getting them attended.
- Participating actively in the mock drill for fire fighting and first aid and guiding the workmen.

b) *Under pinning knowledge*

- Importance and functions of various machines and mechanisms used in spinning preparatory machines.
- Planning maintenance activities and preparing date-wise plans for maintenance and replacement of parts considering their life.
- Role of humidity and temperature in maintaining quality and productivity. Basic knowledge of operating a humidification plant.
- Workloads, work allocation and standard working conditions for maintenance operatives.
- Calculation of maintenance efficiency; time spent for maintenance,

men employed, cost of maintenance, costs of spares consumption, mean time between breakdowns, and the industry norms.

- Factors affecting maintenance.
- Roles and responsibilities of a maintenance supervisor.
- Basic supervisory skills – Listening and Observing, Communicating, Counseling, Taking charge, Reporting and Motivating.
- General management knowledge of managing subordinates, coordinating with workshop, electrical department, stores and production.
- Standing orders and discipline in working.
- Precautions to be taken while working.
- Importance of cleanliness and personal safety.
- Fire fighting and first aid.
- Safety precautions and gadgets to be used in factory.

13.2.5 Supervisor – maintenance – ring frames

a) *Practical competencies*

- Taking round of the work area before the start of the shift and observing the working. Taking information from the previous shift supervisor.
  - Noting down the machines stopped for repairs and arranging for their repairs.
  - Understanding the quality complaints in the machines and arranging for rectifying them.
  - Understanding the works done till now and the works pending in the machines stopped for repairs or for maintenance works or modifications.
- Understanding the production plan and maintenance plan and allocating people for different activities.
- Understanding the machines allotted for various counts and mixings and deciding on the parameters to be checked while doing maintenance like top arm settings, spacers, cots and aprons, spindle tapes, jockey pulley alignment, rings and travelers, traveler clearer setting, etc.
- Verifying the stock of various parts and lubricants and working out the indenting plan and placing indents in time.
- Referring the machinery catalogues and identifying the correct spares needed and ensuring correct spares are put.
- Checking the quality of materials received at stores like spindles, rings, aprons, cots, belts, bearings, top arms, arbours, lappet hooks,

spacers, bobbin holders, separators etc.

- Allocating different tasks to the workers considering their knowledge, skills, maturity and workloads agreed like spindle gauging, top arm setting, bobbin holder maintenance, jockey pulley setting, erection and overhauling etc.
- Checking the maintenance activities in ring frames like routine cleaning, spindle gauging, top arm setting, jockey pulley setting, oiling and greasing etc.
- Working out the changes to be made for change in count or product mix and getting the changes made by the concerned maintenance workers.
- Checking the conditions of machine parts while they are being cleaned/scoured or overhauled and getting the worn-out parts replaced.
- Counseling habitually absenting workers and referring to HRD if needed.
- Counseling / influencing a poor performing worker to produce as per norms.
- Monitoring the stoppages due to breakdowns and analysing the reasons for breakdowns. Preparing action plans and implementing to reduce breakdowns.
- Recording the activities in log book (Report Book) and updating the Machine History book.
- Conducting the tool audits i.e. the tools used for maintenance like spanners, top arm gauge, lubricating and flushing pumps, buffing machines, cot mounting machines, etc.
- Ensuring use of safety gadgets like caps, masks, gloves, aprons and shoes by all maintenance workers.
- Verifying the safety stop motions and getting them attended.
- Participating in mock drill for fire fighting and first aid and guiding the workers.

b) *Under pinning knowledge*

- Importance and functions of various mechanisms used in ring spinning and the maintenance equipments used.
- Planning maintenance activities, preparing date-wise plans for maintenance and replacement of parts as per their life.
- Role of humidity and temperature in maintaining quality and productivity and basic knowledge of operating a humidification plant.
- Workloads, work allocation and standard working conditions for maintenance activities.

- Calculation of maintenance efficiency; mean time between breakdowns, production lost due to breakdowns, the cost of maintenance, time spent for various maintenance activities and the industry norms. Factors affecting maintenance.
- Factors affecting power consumption in spinning, role of maintenance in controlling the power consumption, norms of power consumption per unit production.
- Factor affecting quality of yarn in ring frames and preventive actions.
- Roles and responsibilities of a maintenance supervisor.
- Basic supervisory skills – Listening and Observing, Communicating, Counseling, Taking charge, Reporting and Motivating.
- General management knowledge of managing subordinates, coordinating with workshop, electrical department, stores and production.
- Standing orders and discipline in working.
- Precautions to be taken while doing maintenance works.
- Importance of cleanliness and personal safety.
- Fire fighting and first aid.
- Safety precautions and gadgets to be used in a factory.

13.2.6 Supervisor – maintenance - open-end spinning (rotor spinning)

a) *Practical competencies*

- Taking round of the work area before the start of the shift and observing the working. Taking information from the previous shift supervisor.
  - o Noting down the machines stopped for repairs and the reasons. Arranging for attending to repairs.
  - o Understanding the quality complaints in the machines and arranging for rectification.
  - o Understanding the works done till now and the works pending in the machines stopped for repairs or for maintenance works or modifications.
- Understanding the production plan and preparing maintenance plans and allocating people for different activities.
- Understanding the machines allotted for various counts and mixings and deciding on the parameters to be checked while doing maintenance like opener roller speeds, rotor type and condition, naval size, cots, cheese holders, rotor belts, etc.

- Verifying the stock of various parts and lubricants and working out the indenting plan and placing indents.
- Checking the quality of materials received at stores like rotors, opening rollers, cots, belts, bearings, etc.
- Checking the maintenance activities in rotor spinning machines like routine cleaning, rotor checking and replacement, cheese holder setting, oiling and greasing etc.

b) *Under pinning knowledge*

- Importance and functions of various mechanisms used in rotor spinning machine.
- Planning maintenance activities, preparing date-wise action plans for maintenance and replacement of parts considering their conditions and life.
- Workloads, work allocation and standard working conditions for maintenance operations.
- Calculation of maintenance efficiency; mean time between break downs, cost of maintenance, time spent, production loss due to breakdowns and the industry norms. Factors affecting maintenance.
- Roles and responsibilities of a maintenance supervisor.
- Factors affecting quality of open-end yarn and actions to overcome it.

13.2.7 Supervisor – production - post spinning – winding

a) *Practical competencies*

- Understanding the production plan and allocating the drums/ machines for different activities and monitoring the production on regular basis.
- Understanding the machines allotted for various counts and types and deciding on the changes to be done if needed.
- Verifying the stock of various spinning bobbins and advising the spinning supervisor regarding the productions to be made to complete the orders in each count.
- Verifying the stock of cones in packing or in warping /or the next process and working out the production requirement
- Verifying the count labels in stock and arranging for the labels as needed.
- Verifying the settings of yarn clearers (Mechanical/Electronic) and stop motions and checking their proper functioning.
- Checking the quality of materials; the cone built, the weight

variations, cone hardness, excessive cuts etc., being produced by taking rounds and informing the concerned maintenance personnel for correcting in case of poor quality.

- Checking and controlling the hard wastes generation and cleaning of empty bobbins before sending them to ring spinning.
- Allocating the workers on the machines considering their skills and workloads agreed and the counts working on the machines.
- Checking the knotting/splicing quality of individual drums from time to time.
- Monitoring the pressure of compressed air and its utilization.
- Working out the changes to be made for change in count or product mix and giving instructions.
- Planning the materials needed in advance for a count change and ensuring minimum time for count change.
- Monitoring the humidity and temperature as per requirement by coordinating with the concerned engineering operator.
- Checking the conditions of cone baskets/crates, cone trolleys, empty cones and segregating the bad material.
- Checking the colour codification.

b) *Under pinning knowledge*

- Importance and functions of various mechanisms used in cone winding.
- Importance of yarn clearers, the different systems of yarn clearers and their setting, linking of fault classes to the winding setting.
- Production balancing – Importance and methodology.
- Role of humidity and temperature in maintaining quality and productivity.
- Workloads, work allocation and standard working conditions.
- Calculation of production and efficiency, the industry norms. Factors affecting productivity.
- Factors affecting quality of winding and their remedies.

13.2.8 Supervisor – production – post spinning – doubling-twisting

a) *Practical competencies*

- Understanding the production plan and allocating the machines for different activities.
- Understanding the machines allotted for various counts and types

and deciding on the changes to be done if needed.

- Verifying the stock of various assembly-wound cheeses and advising the winding supervisor regarding the productions to be made to complete the orders in each count.
- Verifying the stock of cones in packing or in warping /or the next process and working out the production requirement
- Verifying the count labels in stock and arranging for the labels as needed.
- Checking the quality of materials; the bobbin/cheese/cone built, the weight variations, cone hardness, etc., being produced by taking rounds and informing the concerned maintenance personnel for correcting in case of poor quality
- Monitoring the functioning of tension capsules and magnets in two-for-one twister.
- Checking and controlling the hard wastes generation and cleaning of exhausted cones/cheeses before sending them to assembly winding.
- Checking the knotting quality of individual knotting heads/ workmen from time to time.
- Monitoring the timely replacement of water in case of wet doubling operations if any.

b) *Under pinning knowledge*

- Importance and function of various mechanisms used in ring doubling and two-for-one twisting machines.
- Factors affecting the quality of a doubled yarn.
- Production balancing – Importance and methodology.
- Role of humidity and temperature in maintaining quality and productivity.
- Workloads, work allocation and standard working conditions.
- Calculation of production and efficiency, the industry norms. Factors affecting productivity.
- Factors affecting the quality of doubled yarns and their remedies.

13.2.9 Supervisor – maintenance – winding

a) *Practical competencies*

- Taking round of the work area before the start of the shift and observing the working. Taking information from the previous shift supervisor.
  - o   Noting down the machines/drums stopped for repairs and

arranging for their repairs.

   ○ Understanding the quality complaints in the machines and arranging for attending them.

   ○ Understanding the works done till now and the works pending in the machines stopped for repairs or for maintenance works or modifications

- Understanding the machines allotted for various counts and mixings and deciding on the parameters to be checked while doing maintenance like drum alignment, tension setting, EYC settings, cone-holder setting, splice or knotting setting, length setting on cones etc.
- Verifying the stock of various parts and lubricants and working out the indenting plan and indenting them in time.
- Referring the machinery catalogue and identifying the required spares and indenting them. Ensuring that correct parts are put on the machines.
- Checking the quality of materials received at stores like drums, prisms for splicing, knotter blades, tension capsules, magazine parts, cone holders, etc.
- Allocating the workers different tasks considering their skills and workloads agreed like setting of cone holders, servicing of knotters and splicers, setting of yarn clearers, setting the machine, etc.
- Checking the maintenance activities in winding section like routine cleaning, drum checking and replacement, cone holder setting, oiling and greasing etc.
- Checking the conditions of machine parts while they are being cleaned/scoured or overhauled

b) *Under pinning knowledge*

- Importance and functions of various mechanisms used in winding (manual as well as automatic) machine.
- Planning maintenance activities and replacement of parts as per their life.
- Role of humidity and temperature in maintaining quality and productivity.
- Workloads, work allocation and standard working conditions.
- Calculation of maintenance efficiency, the cost of maintenance, production loss due to breakdowns, and the industry norms. Factors affecting maintenance.

13.2.10 Supervisor – production – post spinning – reeling

a) *Practical competencies*

- Taking round of the work area before the start of the shift and observing the working. Taking charge from the previous shift supervisor in case of shift working. (Note: Normally as mainly ladies work in reeling section, in number of mills, the reeling works only in general shift).
- Understanding the production plan and allocating the reels for different activities.
- Understanding the machines allotted for various counts and types (British hank or Metric hank – Plain reels or cross reels) and deciding on the changes to be done if needed.
- Verifying the stock of various spinning bobbins or OE cheeses and advising the spinning supervisor regarding the productions to be made to complete the orders in each count.
- Verifying the stock of hanks in bundling section in packing and checking their compatibility for packing (heavy and light hanks)
- Verifying the tie yarn in stock and arranging for them as needed.
- Verifying the conditioning of the cops and their dryness before supplying for reeling.
- Getting the water tank cleaned from time to time as needed in the yarn conditioning area.
- In case of steam conditioning, checking the steam pressure (vacuum pressure and the full working pressure) and the timing of conditioning and monitoring them as per the quality of yarn being reeled.
- Checking the quality of hanks; number of ends, shrinkage and entangle free knots, the weight variations, girth variations, correct tie yarn etc., by taking rounds and getting them corrected as needed.
- Checking and controlling the hard wastes generation and cleaning of empty bobbins before sending them to ring spinning.
- Allocating the workers on the machines considering their skills and workloads agreed and the counts working on the machines.
- Checking the productions of all reelers.
- Checking the knotting quality of individual reeler from time to time.
- Checking the bundling quality and production.
- Checking the bundle weights at random and ensuring that the bundles are of correct weight.
- Checking the quality of dressing the knots.
- Checking the labeling done on bundles.
- Checking the conditions of cane baskets/crates, and segregating the bad material.

b) *Under pinning knowledge*

- Importance of various girth sizes available in reeling.
- Bundling – British system and Metric system.
- Relation between count and the number of knots in a bundle.
- Importance of brushes, and knotting.
- Production balancing – Importance and methodology.
- Workloads, work allocation and standard working conditions.
- Calculation of production and efficiency, the industry norms. Factors affecting productivity.

13.2.11 Supervisor – production – post spinning – yarn packing

a) *Practical competencies*

- Taking round of the work area before the start of the shift and observing the stocks of different yarns in different packing (hanks, cones for bag packing, cones for carton packing, cones for shrink packing etc).
- Understanding the orders to be packed, the production plan and allocating the packers for different activities.
- Checking the stock of packing materials at packing and also at stores and comparing with the requirements depending on the orders to be packed.
- Verifying the quality of packing material received at stores.
- Verifying the weighing balance for its correctness by using standard calibrated weights before starting the work of packing.
- Ensuring that the cones are checked for build and shape, count label, colour code, winder number before putting them in plastic bags.
- Ensuring that correct number of cones is put in each package; cartons, bags, standard pallets for shrink packing as needed.
- Ensuring the bags are tied tightly where bag packing is done.
- Ensuring uniform tension of shrink packing polythene sheet in case of shrink packing.
- Ensuring that hanks are properly dressed and the tie yarns verified before being weighed and kept as per the weight.
- Verifying the bundles for correct weight.
- Checking the bundling quality and bale production.
- Ensuring correct weight and dimensions of bales packed.
- Checking the productions of all packers, recording the package numbers packed by each packer.
- Ensuring that only authorised persons are employed in packing

area, and the names and photos of the packers are displayed prominently as needed.
- Checking the packing quality of individual packer from time to time.
- Maintaining the packing record as per the requirement (legal and statutory).

b) *Under pinning knowledge*

- Importance of packing – Image of the company depends on the packing practices. Poor packing, shortage of materials or mix up in the final packed material can create a very bad impression of the company and customers do not like to make business with such suppliers.
- Different types of yarn packing viz. bag packing, carton packing and pallet packing (Shrink packing) for cones and bale packing for hanks.
- Bundling – British system and Metric system of bale packing.
- Relation between count and the number of knots in a bundle.
- Role of humidity and temperature in maintaining correct weights.
- Workloads, work allocation and standard working conditions for packing.
- Calculation of production and efficiency, the industry norms. Factors affecting productivity.
- Legal and statutory requirements of yarn packing, i.e. the statutory markings, maintaining of packing records, preparing delivery challans, maintenance of stock records, record of people employed in packing and their back ground and contact details.

13.2.12 Supervisor – production – weaving preparation – warping

a) *Practical competencies*

- Taking round of the work area before the start of the shift and observing the stocks of warp beams and different yarns available.
- Understanding the sorts/orders to be beamed.
- Preparing production plan considering the number of ends and set length.
- Allocating the warping machines for different activities.
- Checking the stock of empty beams in stock and comparing with the requirements depending on the orders to be beamed.
- Verifying the condition of beam flanges.

- Verifying the weighing balance for its correctness by using standard calibrated weights before using for weighing.
- Ensuring that the cones are creeled as per the pattern required.
- Ensuring that correct number of cones are creeled
- Ensuring that the tension of each end is set and all yarns have equal tension.
- Ensuring that the ends pass through the stop motion and reed
- Ensuring that the sections are adjusted properly in case of sectional warping
- Ensuring the empty beams are weighed and the weights recorded before starting each beam.
- Ensuring the length set is correct as per the requirement of the beams.
- Working out the beam count for each beam by weighing the full beams.
- Checking the productions of all warping machines.
- Maintaining the warping record as per the requirement.

b) *Under pinning knowledge*

- Importance of warping and functions of a warping machine.
- Different types of warping viz. beam warping and sectional warping.
- Various mechanisms in warping machines and the controls available.
- Relation between cone weight, count and the yarn content in the cone.
- Workloads, work allocation and standard working conditions.
- Calculation of production and efficiency, the industry norms. Factors affecting productivity.
- Safety, legal and statutory requirements applicable to warping operations.

13.2.13 Supervisor – production – weaving preparation – sizing

a) *Practical competencies*

- Taking round of the work area before the start of the shift and observing the stocks of sized beams and pending requirement.
- Understanding the sorts/orders to be beamed.
- Preparing production plan considering the number of number of warp beams and set length.

- Allocating the sizing machines for different activities.
- Checking the stock of empty beams and comparing with the requirements depending on the orders to be beamed.
- Verifying the condition of beam flanges and getting them corrected.
- Verifying the weighing balance for its correctness by using standard calibrated weights before using for weighing.
- Ensuring that the warp beams are creeled as per the pattern required.
- Working out the size recipe and getting materials from stores as needed.
- Ensuring that the size cooking is done as per the size recipe suggested considering the type of yarn and the number of ends.
- Ensuring the size ingredients are stored and handled in a safe way.
- Ensuring the sow box is washed well before starting a new set.
- Ensuring that the tension is set and stretch is maintained as minimum as possible and uniformly.
- Ensuring that the steam condensates are discharged from time to time.
- Ensuring that the ends pass through the stop motion and reed.
- Ensuring that the steam pressure and the temperature are maintained
- Ensuring that drying is uniform and adequate.
- Ensuring that the empty beams are weighed and the weights recorded before starting each beam.
- Ensuring the length set is correct as per the requirement of the beams.
- Working out the size pick up for each beam by weighing the full beams.
- Checking the productions of all sizing machines.
- Maintaining the Sizing record as per the requirement.

b) *Under pinning knowledge*

- Importance of sizing.
- Different types of sizing machines viz. two cylinder slasher sizing, multi-cylinder sizing, infra red drying in sizing, hot air drying in sizing and single end sizing.
- Importance and functions of various mechanisms in sizing machines.
- Working out size pick up and methods of controlling size pick up.
- Role of moisture content in maintaining correct weights.

- Workloads, work allocation and standard working conditions.
- Calculation of production and efficiency, the industry norms. Factors affecting productivity.
- Safety, legal and statutory requirements applicable to sizing operations.

13.2.14 Supervisor – production — weaving preparation – weft winding

a) *Practical competencies*

- Taking round of the weaving work area before the start of the shift and observing the stocks of pirns prepared and pending requirement.
- Understanding the sorts/orders being run and the requirement of weft yarn.
- Preparing production plan considering the number of winding units to be allotted for different count.
- Checking the stock of empty pirns in stock and arranging for getting the required quantity back from loom shed.
- Ensuring that the cones received for weft winding are of correct count and specifications.
- Ensuring that the winding is done with proper tension and with specified length of reserve yarn and cop bottom.
- Ensuring that the tension is set and stretch is maintained as needed uniformly.
- Checking the productions of all winding machines.
- Maintaining the weft winding record as per the requirement.

b) *Under pinning knowledge*

- Importance of pirn winding.
- Different types of pirn winding machines viz. single spindle charaka, multi spindle manual pirn winding and automatic pirn winding machines.
- Various mechanisms in pirn winding machines.
- Adjusting the chase length depending on the yarn count.
- Role of moisture content in maintaining correct shape and tightness of cops.
- Workloads, work allocation and standard working conditions.
- Calculation of production and efficiency, the industry norms. Factors affecting productivity.

13.2.15 Supervisor – production – weaving – shuttle looms

a) *Practical competencies*

- Taking round of the work area before the start of the shift and observing the stocks of sized beams and pending requirement.
- Understanding the sorts/orders to be beamed.
- Preparing production plan considering the orders being processed, the looms working and the weavers beam in stock.
- Ensuring that the weaver's beams are drawn as per the pattern required.
- Arranging for gaiting the looms as per programme.
- Ensuring that the tension is set and maintained uniformly.
- Ensuring the length set is correct as per the requirement of the fabric.
- Inspecting the quality of the first piece after gaiting a loom.
- Checking the productions of all weaving machines from time to time.
  Maintaining the weaving production record as per the requirement.

b) *Under pinning knowledge*

- Weaving basics, types of weaves.
- Different types of Shuttle loom viz. over pick looms, under pick looms, pirn changing auto looms, shuttle changing auto looms, box changing looms, dobby looms and Jacquard looms.
- Various mechanisms in shuttle looms and their functions.
- Factors affecting quality and productivity in shuttle looms.
- Role of humidity and temperature in getting weaving efficiency. Basic knowledge of operating the humidification plant to achieve the required humidity.
- Workloads, work allocation and standard working conditions.
- Calculation of production and efficiency, the industry norms. Factors affecting productivity.
- Safety, legal and statutory requirements applicable to weaving operations.

13.2.16 Supervisor – production – weaving – shuttle – less Looms

a) *Practical competencies*

- Taking round of the work area before the start of the shift and

observing the stocks of sized beams and pending requirement.
- Understanding the sorts/orders to be beamed.
- Preparing production plan considering the orders in hand, the looms working and the weavers beam in stock.
- Ensuring that the weaver's beams are drawn and gaited as per the pattern required.
- Ensuring that the tension is set and maintained uniformly.
- Ensure the length set is correct as per the requirement of the fabric.
- Inspecting the first piece after gaiting a loom.
- Checking the productions of all weaving machines from time to time.
  Maintaining the weaving production record as per the requirement.

b) *Under pinning knowledge (Theory)*

- Weaving basics, types of weaves.
- Different types of shuttle-less looms viz. gripper looms, rapier looms, air-jet looms, dobby looms and jacquard looms.
- Various mechanisms in shuttle-less looms i.e. weft feeding mechanisms, selvedge mechanisms, cutters, etc.
- Factors affecting the quality and productivity in a shuttle-less loom.
- Role of humidity and temperature in getting weaving efficiency. Basic knowledge of operating humidification plant to maintain the humidity.
- Workloads, work allocation and standard working conditions.
- Calculation of production and efficiency, the industry norms. Factors affecting productivity.
  Safety, legal and statutory requirements applicable to weaving operations.

13.2.17 Supervisor – production – wet processing – yarn dyeing in package form

a) *Practical competencies*

- Taking round of the work area before the start of the shift and observing the machines working on different lots and the lots to be taken next for dyeing.
- Understanding the shades to be dyed, the shade matching, the group of dyes and the dyeing recipe and arranging the dyes and chemicals as needed.
- Checking the stock of soft wound yarn packages and advising the winding to produce the soft packages.

- Checking the dye recipe prepared for the correctness of component chemicals.
- Checking the quality of water before using it for dyeing.
- Checking for the proper cleaning of vessels before taking the lots for dyeing.
- Checking the package density of soft wound packages.
- Getting the packages loaded in the carriages and ensuring that packages are not touching each other.
- Programming the dyeing process cycle as per the lot process card decided after shade matching.
- Ensuring that the packages are taken out after the process cycle of dyeing and washing are completed and sent for drying.
- Checking the shade after drying at inner and outer portion of cheese and advising for cutting at final winding.
  Ensuring that dyed yarn is wound on cones on properly segregated winding machines, and the winding machines are thoroughly cleaned before taking for winding.

b) *Underpinning knowledge*

- Principles and practices of yarn dyeing in package form.
- Importance of various features of a HTHP yarn dyeing machine.
- Density of soft yarn package and its importance in getting uniform shade.
- Various dyes used for dyeing and dyeing methods.
- Factors affecting quality of dyeing and precautions to be taken while dyeing cheeses in package form.
- Testing of dyes and chemicals for their purity and concentration. Testing dyed yarn for fastness properties.

13.2.18 Supervisor – production – wet processing – kier boiling

a) *Practical competencies*

- Taking round of the work area before the start of the shift and observing the kiers working on different lots and the lots to be taken next for boiling.
- Checking the stock of material ready for kier boiling.
- Ensuring proper weighment of material before loading into kiers.
- Checking the quality of water before using it for kier boiling.
- Checking quantity of caustic soda and soda ash mixed in the kitchen and the pH of kier solution.

- Checking the kier boiling parameters set like temperature, pressure, and time etc before starting the machine.
- Ensuring the kier is cleaned thoroughly before starting a new lot for boiling.
- Checking the condition of hoist and containers before starting the work.
- Ensuring the use of safety gadgets like helmet, mask, gloves and gum shoes by the workers.
- Ensuring thorough cleaning of chemical containers before disposing
- Maintaining the records of production and chemical consumption.

b) *Underpinning knowledge*

- Purpose and Process of Kier boiling
- Functions of various controls in a kier
- Importance of liquor ratio and cost of kiering
- Precautions to be taken while kier boiling for avoiding stains
- Controlling water pollution
- Precautions to be taken while storing chemicals

13.2.19 Supervisor – production – garment manufacture – sewing

a) *Practical competencies*

- Taking round of the work area and getting details of styles and P.O being worked, the garments stitched till now and to be stitched.
- Understanding the tech pack of the style with details of size, stitches and seams and explaining to tailors.
- Ensuring that tailors are engaged as per their skill.
- Ensuring correct needles and sewing threads are given to each machine.
- Working with the IE, QC and mechanics while setting a batch and ensuring speedy batch setting.
- Checking the stitch quality of each tailor and the machines and getting them attended if needed.
- Monitoring the production of each tailor and recording production of line on hourly basis.

b) *Underpinning knowledge*

- Knowledge about different stitches, seams and sewing machines.

- Basic requirements of garments, the stitch allowances, aligning grin.
- Precautions to be taken before starting a stitching operation. Understanding the SAM value and working out the production expected.

13.2.20 Supervisor – garment finishing and packing

a) *Practical competencies*

- Taking round of the work area and getting details of garments in each style ready for packing, garments in each style received for finishing, and the garments to be received in each style from sewing.
- Verifying the measurements, folding and packing details and arranging for workers for the jobs.
- Arranging people for different finishing activities like quality inspection, stain removing, washing, trimming, ironing, folding, metal detecting, packing in bags and cartons etc.
- Verifying the people engaged for packing and ensuring that they are all in the approved list of packers and their photos are displayed in the entrance of packing section.
- Checking the quality of packing materials and other accessories received like hangers, polybags, card boards, single garment boxes, cartons, stickers, wash care labels and other packing accessories like pins, clips, twine etc.
- Keeping account of garments finished and packed in each style and size.
- Storing the packed material in designated place.
- Offering the packed garments for final inspection by the representative of buyer or the final statistical audit team of the company.
- Supervising the loading operations.
- Ensuring proper housekeeping in finishing and packing area.

b) *Underpinning knowledge*

- Importance of packing and specific packing requirements of the customers.
- Importance and process of ironing, specific ironing requirements for different fibres and fabric qualities and garments.
- Washing procedure for garments specific to the materials used in garments.

- Packing regulations, C-TPAT compliance.
- Statutory requirement of records, delivery challans and invoice generation.

The above are some examples and the concerned mill or the garment factory should write the competency needed and the underpinning knowledge required depending on the type of machinery, the processes involved and the product quality requirement.

14

Control points and check points

It is essential for a supervisor to have clarity on the points to be controlled to achieve the targets and those to be checked to ensure the process in control. Some examples of control points and check points are discussed in this chapter. However, it is suggested that these points need to be reviewed from time to time and modified to suit the requirements of individual companies and their targets. Also it is suggested that each mill or a garment factory prepare their own "control points and check points" and display them in the work area so that the supervisors shall refer and follow.

14.1   Process spinning

In spinning there are different sub processes like mixing, blow room, carding, drawing, combing, speed frames and ring frames or rotor spinning. Let us discuss on the control points and check points in these processes.

14.1.1 Mixing

Control points

- Selection of bales considering the parameters like length, strength, fineness, colour, trash, maturity, neps, etc., and their variations to meet the quality requirements of the yarn proposed.

- Deciding the proportion of different components in the mixing considering their properties, cost, age and stocks.
- Deciding the quantity of mixing to be done at a time.
- Issuing mixing in time.
- Thorough opening and homogeneous mixing.
- Adding prescribed proportion of useable soft wastes in the mixing.
- Deciding the addition of spin-finish, hygroscopic and antistatic agents, tinting colours, etc depending on the materials being used.
- Adequate conditioning of mixing before feeding to the Blow room.
- The work allocation for employees.
- Disposal of bale packing materials like bale cover cloth, bale hoops or wires.

Check points

- Whether the lot numbers of the bales received are as per the plan?
- Whether the floor is cleaned properly before laying the mixing?
- Whether the bales are kept in their designated place as per mixing floor plan?
- Whether bales are opened gently?
- Whether the bale cover cloth and bale hoops are removed and kept at designated place properly?
- Whether surface of bales are cleaned before opening?
- Whether cotton is opened and broken into smallest tufts as explained?
- Whether the contaminations are removed by proper checking?
- Whether the antistatic agent / spin finish / tint etc are properly prepared as per plan and applied uniformly?
- Whether the layers of opened material are put properly in the stack to have a homogeneous mixing?
- Whether the identification board is updated?
- Whether the temperature and humidity are as per the requirement?
- Whether the required time is allowed for conditioning?
- Whether the men employed are as per plan?
- Whether the required quantity of mixing is prepared?
- Whether the usable wastes are received with proper identification?
- Whether the wastes are properly checked and cleaned before adding in the mixing?
- Whether the quantity of useable wastes added is maintained uniformly in all mixings?
- Whether the workers are using safety gadgets like masks and headgear while working?

14.1.2 Blow room

Control Points

- Selection of opening and cleaning points and their sequence as per the mixing.
- Selection of process parameters like speed, setting, hank of material delivered, length of lap, the calendar and rack pressure in scutchers, etc.
- Engaging trained workmen.
- Evolving and implementation of maintenance schedules.
- Providing and maintaining safety gadgets as required.
- Deciding the work allocation per employee.
- Deciding the frequency and systems for waste evacuation and transportation.
- Deciding suitable identification systems for wastes and good material delivered.

Check points

- Whether all the cleaning and opening points required are in working order?
- Whether the bypasses are done as per plan?
- Whether the synchronisation of working of machines is suiting the quality requirement?
- Whether speeds of the beaters, fans, feed rollers etc., are as per plan?
- Whether the settings are as decided?
- Whether the lines are cleaned as per plans, and before changing the mixings?
- Check the droppings for presence of good cottons.

○ Whether the hank of lap or the weight per meter of delivered material sheet are as per plan and are uniform?
○ Whether the workmen follow safety regulations regularly? No materials should be kept blocking access to electrical control panels and fire extinguishers. All fire extinguishers should be in working position and in their allocated places. The fire exit doors should be free and the passages are not blocked.
○ Whether neps generation and fibre rupture are within control?
○ Whether the men employed are adequately trained?
○ Whether wastes are removed in time and labelled properly?
○ Whether the temperature and humidity are as per requirement?

14.1.3 Carding

Control points

■ Selection of process parameters, viz, card clothing, speeds, settings, drafts and hank.
■ Preparing and maintaining the schedules for preventive maintenance like setting, grinding, mounting, etc.
■ Engaging trained workmen.
■ Maintaining the required temperature and humidity.
■ Designing and providing safety gadgets.
■ Following suitable colour codification and channelisation.
■ Deciding the work allocation per employee.
■ Deciding frequency and systems for waste evacuation and their disposal and implementing them.

Check points

○ Whether the cards are allocated to different mixings and hanks as per the plan?

○ Whether the wire points on the cards match with the requirement of the mixing?

○ Whether the machines are in good condition?

○ Whether card wire points are sharp and clean?

○ Whether the settings are done as specified?

○ Whether the wheels put are as per requirement?

○ Whether the quality of web is good without neps, holes and ragged or bunched selvedges?

○ The breakages and their reasons.

○ Whether the wastes are removed in time?

○ Whether the temperature and humidity are as per requirement?

○ Whether the laps fed are of good quality and without licking?

○ Whether the hank of sliver produced is as per plan?

○ Whether the sliver is uniform and U% is within norms?

○ Whether the tenters carry out the work as specified?

○ Whether the machines and the materials are labelled properly for identification?

○ Whether all stop motions are working properly?

○ Whether the production obtained is meeting the targets?

○ Whether the card is giving the cleaning efficiency as required?

○ Whether the neps removal efficiency is as per requirement?

○ Whether the trash and neps in sliver are within control?

○ Whether the fibre rupture is within tolerable limits?

○ Whether the cans and springs used are of required quality?

○ Whether the wastes are removed from spring bottom before feeding cans to cards?

14.1.4 Drawing

Control points

- Deciding the process parameters, viz settings, draft, number of doublings, speed, hank, size of trumpet, length of sliver in can, pressure on the drafting rollers, type of cots etc.
- Deciding the colour codification and channelisation.
- Engaging trained workmen.
- Evolving and implementing maintenance schedules.
- Deriving and providing required humidity and temperature.
- Deciding the work allocation to the employees.

Check points

- Whether the settings are as per requirement?
- Whether the conditions of the machine parts like bottom drafting rollers, top roller cots, end bushes, saddles, hosepipes, springs, belts, bearings, etc., are good?
- Whether the cans and springs are in good condition?
- Whether the wheels are put as per calculations?
- Whether the hank of sliver produced is as per requirement and the variations are within limit?
- Whether the machine is running at the planned speed and giving the required production?
- Whether all the stop motions are functioning properly and without delay?
- Whether there is any overlapping of slivers in the feed?
- Whether the voltage variations are within control for auto leveller draw frames?
- Whether the sliver test A% is within norms for auto levelled material?
- Whether the colour codification and channelisation are followed as per plan?
- Whether the workmen employed are adequately trained?
- Whether the quality of cans and springs are as specified?
- Whether the cans are properly cleaned before putting in the machine?
- Whether the coiling is proper?
- Whether the trumpets are of correct size and have smooth inner surface?
- Whether the surface of the sliver passage in the creel is smooth?
- Whether the scanning rollers adopted are of correct size?
- Whether the temperature and humidity are maintained as per plan?

14.1.5 Combing

Control points

- Selection of process parameters, viz: feeding lap hank, half-laps, speeds, settings, drafts and delivery hank, percentage of noil to be extracted, maximum fibre length in noil, etc.
- Deciding and maintaining the schedules for preventive maintenance like half-lap mounting, setting, brush mounting, buffing etc.
- Engaging trained workmen.
- Maintaining the required temperature and humidity.
- Designing and providing safety gadgets.
- Deciding and following suitable colour codification and channelisation.
- Deciding the work allocation for employees.
- Deciding frequency and systems for waste evacuation and their disposal and implementing them.

Check points

- Whether the mixing and hanks running are as per plan?
- Whether the machines are in good condition?
- Whether half-lap points are sharp and clean?
- Whether the settings are done as specified?
- Whether the wheels put are as per requirement?
- Whether the quality of web is good without piecing marks?
- Whether the breakages are in control and what are the reasons for breaks?
- Whether the wastes [Noil] are removed in time in case of manual waste removing system?

○ Whether the temperature and humidity are as per requirement?
○ Verify whether laps fed are of good quality and there is no licking.
○ The hank of sliver produced is as per plan.
○ The sliver is uniform and U% is within norms.
○ Whether the tenters carry out the work as specified?
○ Whether the machines and the materials are labelled properly for identification?
○ Whether all stop motions are working properly?
○ Whether the production obtained is meeting the targets?
○ Whether the neps removal efficiency is as per requirement?
○ Whether the neps in sliver are within control?
○ Whether the cans and springs used are of required quality?
○ Whether the wastes are removed from spring bottom before feeding cans to cards?

14.1.6 Speed frame

Control points

■ Deciding the process parameters, viz settings, speeds, draft, condensers and spacers, top arm pressure, hank of rove, twist multiplier.
■ Deciding package parameters like coils per inch, taper, lift and diameter, weight etc.
■ Deciding on colour codification and channelisation.
■ Engaging trained workmen.
■ Evolving maintenance schedules and implementing.
■ Deciding and providing required temperature and humidity.
■ Deciding the work allocation for employees.

Check points

○ Whether the hank of rove is as per plan and the variations are within norms?

○ Whether the stretch is in control in all the spindles?
○ Whether the breakages are within control?
○ Whether the U% of the rove is within limits?
○ Whether the machines are running at specified speeds and giving the required production?
○ Whether all the parts of the machines are in good condition?
○ Whether the settings and alignments of rollers in drafting zone are proper?
○ Whether the drafting zone is kept clean?
○ Whether the settings are as per plan?
○ Whether the top arm pressures are as required and uniform on all spindles?
○ Whether the spacers, condensers, sliver guides, false twisters etc., are as per plan?
○ Whether the workers are following the work practices as specified?
○ Whether the workmen are adequately trained?
○ Whether the colour codification and channelisation are followed as per plan?
○ Whether the maintenance is carried out as per plan?
○ Whether the transportation of bobbins to ring frames is as per plans?
○ Whether the temperature and humidity are maintained as required?
○ Whether the house keeping is as required?
○ Whether sufficient empty bobbins are available?
○ Whether bobbins are cleaned properly before putting on the machine?

14.1.7 Ring frame

Control points

- Deciding and adopting process parameters, viz. count, twist, speeds, settings, draft, rings, travellers, spacers, chase, winding and binding ratio etc.
- Engaging sufficient and trained workmen.
- Deciding and providing required air changes, humidity and temperature.
- Deciding colour codification and implementing.
- Deciding the work allocation for employees.
- Evolving the maintenance schedules and activities and implementing them.

Check points

- Whether the ring frames are working as per the count pattern decided?
- Whether the Bobbins fed are as per the plan?
- Whether the count of yarn is as per requirement and the variation is within limit?
- Whether the TPM of yarn is as per requirement and the variation is in control?
- Whether the machine speeds are as per plan?
- Whether the required production and efficiency is obtained as per plan?
- Whether the machine settings are as per the parameters planned?
- Whether the condition of the machine parts is good?
- Whether the yarn uniformity, appearance, hairiness etc., are as per requirement?
- Whether the breakages are in control?
- Whether the men allocation is as per norms?
- Whether the workmen are adequately trained?
- Whether the maintenance is done as per plans?
- Whether the bobbins used are cleaned properly before using on the machine?
- Whether the pneumafil wastes are removed in time as per schedule?
- Whether the colour codification and channelisation are followed properly as per plan?
- Whether the wastes generated are within limits?
- Whether the wastes are disposed with proper accounting and labelling?
- Whether the drafting zone and yarn path are kept clean?
- Whether the temperature and humidity are maintained as per need?

14.2   Process – post spinning

In post spinning section normally winding, reeling and yarn packing operations are considered.

14.2.1 Cone winding

Control points

- Selection of suitable process parameter considering the customer requirements: The parameters are speeds, yarn clearer settings, tension, cone dimensions, wax quality and settings at contamination channel.
- Cone identifications: viz. Cone tip, Cone label, Winder No. Machine No, Lot No. Contract No.
- Engagement of trained workmen.
- Evolving maintenance schedules and activities, and implementing them.
- Deciding the work allocation and production targets.

Check points

- ○ Clearing efficiency of yarn clearers.
- ○ Production efficiency.
- ○ Winding drum-wise performance.
- ○ Winder-wise production and wastes generated.
- ○ Whether the paper / plastic cones used are of required quality?

○ Cone quality.
○ Weight variations between cones.
○ Hard wastes generated and the reasons.
○ Splicing quality.
○ Friction value for waxed yarns.
○ Cone identification.
○ Work practices.
○ Cleanliness and house keeping.
○ Condition of machine parts.
○ Cone hardness.
○ Increase in hairiness and imperfections after winding.
○ Working of all stop motions.
○ Whether the spinning bobbins have the correct identification as decided?
○ Whether the workers are using the spinning doffs with the machine traceability in view?
○ Spinning doff stocks and winder allocation.
○ Cones required in each lot to complete packing.

14.2.2 Reeling

Control points

■ The count to be reeled.
■ The girth and length.
■ The tie yarn to be used.
■ The quantity to be reeled.
■ Number of reels to be allotted for different counts.
■ Type of reel to be used – Cross or Plain.
■ Number of hanks per bundle.
■ Bundle weight and Bale weight.

- The frequency of changing water in case of water conditioning.
- Production norms as per the count.

Check points

- Whether the yarn is adequately conditioned?
- Whether the yarns are dried properly?
- Is there any stain on the feed yarn packages?
- Whether the swifts are opened fully and the length of yarn wound per revolution is as per norms?
- Whether the yarns are working with least possible breaks?
- Whether each reeler is giving the required production?
- Whether the hanks are dressed properly without entanglements?
- Whether the weights of the hanks are matching to the count and length of hank?
- Whether the bundle weights are as per the weight printed on the bundle?

14.3   Process – doubling

In doubling section there are sub processes like assembly winding, ring doubling or twisting on 'two-for-one' twister. The ring doubling may be dry doubling or wet doubling; whereas in two-for-one twisting there is no provision for wet doubling. Let us discuss on control points and check points.

14.3.1 Assembly winding

Control points

- Selection of process parameters: speeds, tension, feed package positions, delivery package dimensions, etc.
- Assembly wound package identifications.
- Allocation of trained workmen.
- Allocation of number of drums to winders and fixing production norms.
- Planning maintenance operations and implementing them.

Check points

- Uniformity in tension of the components of assembled package.
- Functioning of stop motions.
- Identification of packages.
- Stock of assembly wound packages and production required for completing the lot.
- Whether the breakages are in control?
- Weight variation between delivery packages.
- Whether the winders are controlling the tension of each component while attending to breakages?
- Cleanliness and house keeping.
- Whether the feed packages are creeled as per the Feed Package Position Plan?
- Whether the tension weights used are uniform on all drums and are as per plan?
- Whether the delivery packages are having required package density?
- Whether the yarn path is free from rough surfaces, serrations, and groves?
- Whether the machine is giving production as per plan?
- Whether the package storing area is clean?
- Whether the workers are aware of the material they are processing and the quality requirements?
- Whether the workers are allotted as per the work norms?
- Whether the workers are adequately trained?
- Whether the maintenance activities are carried out as per plan?

14.3.2 Doubling (Twisting)

Control points

- Selection of process parameters: twist per inch, speeds, tension, feed package dimensions and delivery package dimensions.
- Travellers in case ring doublers.

- Delivery package identifications.
- Allocation of trained workmen.
- Allocation of number of drums/ spindles to tenters and fixing production norms.
- Planning maintenance operations and implementing them.

Check points

- Uniformity in tension in each spindle.
- Whether the twist introduced is as per requirement and the variations are within norm?
- Functioning of stop motions.
- Identification of packages.
- Stock of assembled packages and production required for completing the lot.
- Stock of twisted materials and the production required for completing the lot.
- Whether the breakages are in control?
- Weight variation between delivery packages.
- Whether the tenters are controlling the tension of each component while attending to breakages in case of cabled yarns?
- Cleanliness and house keeping.
- Whether the tension is maintained uniform on all drums/spindles and is as per plan?
- Whether the delivery packages are having required package density?
- Whether the yarn path is free from rough surfaces, serrations, and groves?

○ Whether the machine is giving production as per plan?
○ Whether the package storing area is clean?
○ Whether the workers are aware of the material they are processing and the quality requirements?
○ Whether the workers are allotted as per the work norms?
○ Whether the workers are adequately trained?
○ Whether the maintenance activities are carried out as per plan?
○ Whether the water is replaced in time in case of wet doublers?
○ Whether the rings are lubricated properly with correct oil in case of wet doubling?

14.4   Process – weaving preparatory

In weaving preparatory there are sub processes of warping, sizing, pirn winding in case of shuttle looms, and beam drawing. The warping may be of beam warping or sectional warping depending on the situation. Let us discuss on control points and check points.

14.4.1 Warping

Control points

■ Selection of process parameters like number of ends in the creel, length of yarn in each beam, warping speed, number of beams per set, tension, creeling position of feed packages, number of sections and leasing plan in case of sectional warping.
■ Beam identification.
■ Engagement of skilled employees.

- Deciding of work norms and allocation of workmen.
- Designing maintenance activities and implementing them.

Check points

- ○ Whether the cones are creeled as per plan?
- ○ Whether the feed material is as per plan?
- ○ Whether the empty beams are in good condition?
- ○ Whether the stop motions and brake are functioning properly?
- ○ Whether the length set is as per plan?
- ○ Whether the speed is maintained as per plan?
- ○ Whether yarn tension is maintained uniformly on all ends, and is as per plan?
- ○ Whether the breakages are within norms?
- ○ Whether the machine is giving the required production and efficiency?
- ○ Whether the beam cards are entered with all relevant information?
- ○ Whether the ceramic guides and tension pins in yarn path are in good condition?
- ○ Whether the fans on the creel are properly functioning?
- ○ Whether the workers are following the material handling systems as per requirements?
- ○ Whether the tenters are putting the correct type of knot while mending breaks?
- ○ Whether the beam weights are uniform and as per calculation?
- ○ Whether the humidity is maintained as per requirement?
- ○ Whether the hard wastes generated are within the norms?

14.4.2 Sizing

Control points

- Selection of process parameters like number of beams in the creel, total number of ends in weaver's beam, length of yarn in each beam, sizing speed, number of beams per set, tension, creeling position of feed packages, size recipe, size pick up, temperature of sow box and drying, squeezing pressure, steam pressure, beam width, leasing system, number of drying cylinders to work and number of sow boxes i.e. single or double.
- Beam identification.
- Engagement of skilled employees.
- Deciding of work norms and allocation of workmen.
- Designing maintenance activities and implementing them.
- Collection of wastewater and treating them.
- Collection of water vapours and wastes and disposing.

Check points

- Whether the warper beams are creeled as per plan?
- Whether the empty beams are in good condition?
- Whether the stop motions and brake are functioning properly?
- Whether the length set is as per plan?
- Whether the speed is maintained as per plan?
- Whether tension and stretch are maintained as per plan?
- Whether the machine is giving the required production and efficiency?
- Whether the beam cards are entered with all relevant information?
- Whether the reed, expansion comb and drop pins are in good condition?
- Whether the size is prepared as per plan and the viscosity is as per requirement?
- Whether the size pick up is uniform and as per plan?
- Whether the denting is as per plan?
- Whether the moisture content in the delivered sized yarn is as per requirement?
- Whether the sized yarn has required strength and elongation?
- Whether the condition of the squeezing rollers is good?
- Whether the squeezing pressure is uniform and as per requirement?
- Whether the steam and water consumption are as per norms?
- Whether the wastewater is collected properly and sent for treatment as per norms?
- Whether the steam and water vapour are collected and disposed as planned.

○ Whether the workers are following the material handling systems as per requirements?

○ Whether the tenters are putting the correct type of knot while mending breaks?

○ Whether the beam weights are uniform and as per calculation?

14.5   Process – weaving

In weaving department there may be different types of looms like plain power looms, automatic looms with shuttles, shuttle-less looms with different weft feeding mechanisms like projectile, rapier, air-jet, water-jet and so on. The looms might have different attachments like dobby, jacquard, double beam feeding, and pick at will and so on. Therefore it is necessary for the mill people to decide on the controls and checks. Here some general guidelines are given for control and checks.

14.5.1 Looms

Control points

■ Selection of process parameters like speed, weave, healds and reed, number of shafts, ends and picks per inch, grey width, timings, settings of various motions, type of emery roller, type of temples, length of piece to be doffed and so on.

■ Fabric identification.

■ Evolving work norms and work allocation.

■ Engagement of trained workmen.

■ Deciding and maintaining required temperature and humidity.

Check points

○ Gaiting of correct weavers beam on looms.

- ○ Loom speed and efficiency.
- ○ Fabric defects on loom.
- ○ Grey width and its uniformity.
- ○ First piece inspection after gaiting.
- ○ Fabric cover.
- ○ Selvedge condition.
- ○ The generation of fents, rags and chindis and their reasons.
- ○ The relative humidity and temperature in the shed.
- ○ Stoppages for various reasons.
- ○ Condition of reeds and healds.
- ○ Let-off beam tension.
- ○ Condition of emery rollers.
- ○ Whether the workmen are adequately trained?
- ○ Whether the number of persons engaged is as per plan?
- ○ Whether the maintenance is carried out as per schedule?
- ○ Whether the doffed fabrics are marked for identification as per plan?
- ○ Whether the safety gadgets are used by workmen as required?

14.6   Process – wet processing

Wet processing has number of sub processes like desizing, singeing of yarns and fabrics, scouring, bleaching of yarns and fabrics, mercerising of yarns and fabrics, dyeing of yarns and fabrics, printing, finishing, inspection and folding and so on. The controls and checks depend mainly on the process and the chemicals being used. However, general guidelines for some of the sub processes are given herewith.

14.6.1 Yarn singeing

Control points

■ Deciding the process parameters depending on the raw material and count, voltage to be set in case of electrical singeing, speed, height of flame in case of gas singeing, delivery package dimensions.
■ Providing of skilled employees.
■ Deciding on work norms and allotting men.
■ Providing the required safety gadgets and equipments.

Check points

○ Whether the winding speed is as per plan?
○ Whether the voltage/flame height is uniform and as per plan on all singeing elements?
○ Whether the singed yarns have uniform shade within packages and between packages?
○ Whether all the singeing elements are giving the same performance efficiency?
○ Whether the reduction in hairiness is as per planned?
○ Whether the workers employed are adequately trained?
○ Whether the safety gadgets are in order, and being used by all the employees?
○ Whether the singed materials are properly marked as per plans?
○ Whether the shaded materials are segregated from good material as per plan?
○ Whether the gas cylinders are stored as per the safety regulations?
○ Whether the production is as per planned?
○ Whether the fire extinguishers are in order and kept at designated place?

14.6.2 Yarn mercerising

Control points

- Deciding and selection of process parameters, viz: machine setting, mercerising time and temperature, stretch, time and temperature, caustic lye concentration, washing and neutralising sequences, batch size, and liquor ratio.
- Deciding acceptance criteria for degree of mercerisation, quality of raw materials and chemicals.
- Employing trained and qualified employees.
- Evolving production norms.
- Evolving norms for consumption of chemicals, water and steam.

Check points

- Whether the raw materials, chemicals, and auxiliaries are checked and ensured for meeting the acceptance criteria?
- Whether the concentration of caustic lye is as required?
- Whether the temperature of the caustic lye is maintained as needed?
- Whether the mercerising process cycle time is maintained as scheduled?
- Whether the hanks are with the girth and weight as specified?
- Whether the stretch is maintained as planned?
- Whether the neutralising sequence is followed as per programme?
- Whether pH of neutralising liquor is maintained as specified?
- Whether Barium Activity Number of the mercerised material is as per norms?
- Whether the production is achieved as per norms?
- Whether the men employed are adequately trained?
- Whether the consumption of chemicals, steam and water are as per plan?

14.6.3 Yarn dyeing (HTHP)

Control points

- Raw material quality.
- Preparation of dye package, selection of dye tubes or springs, winding machine parameters viz speed, package dimensions, package density.
- No of cones to be loaded per lot and loading of carrier.
- Water quality, the softness, turbidity and purity.
- Selection of dyes, chemicals and auxiliaries.
- Dye recipe preparation.
  - Use of colour kitchen.
  - Providing the required containers, balances and safety equipments.

- Deciding the dyeing operation sequence, time, temperature, steam pressure, liquor ratio, number of cycles, number of washes, etc.
- Determination of after treatment depending on the dye and depth.
- Determination of acceptance criteria for fastness, friction values, shade variation between and within lots, whiteness index in case of bleaching, and the physical properties of final yarn.
- Employing trained employees.
- Deciding parameters for final winding, the shade to be cut, final cone weight, the wax quality, winding speed and the cone angle.

Check points

- ○ Grey yarn shade uniformity.
- ○ Whether the sample gives uniform dyeing?
- ○ Quality of soft wound package i.e. density, diameter and weight.
- ○ Quality of dye tube.
- ○ Condition of winding machine.
- ○ Winding quality.
- ○ Suitability of hoist for the load.
- ○ Quality of spindles in the carrier.
- ○ Space between spindles and dye tube.
- ○ Dye package compressor functioning.
- ○ Hardness, pH and turbidity of water used.
- ○ Quality of chemicals, dyes and their expiry period.
- ○ Availability of alternatives and combinations for dyes and chemicals.
- ○ Weight and measurement of dyes and chemicals.
- ○ Preparation of dye paste/liquor as per laid out procedure.

○ Cleanliness of containers before use.
○ Correctness of balances and measures.
○ Completeness of process sheet.
○ Calibration of steam and pressure gauges.
○ Percentage utilisation of dye vessel.
○ Fastness as per requirement.
○ Pressure in dyeing machine out-in and in-out flow.
○ Intermediate additions as per time.
○ Rise in temperature and cooling time.
○ Condition of pH during dyeing process.
○ Dyeing and fixation time and temperature.
○ Finishing temperature and time.
○ Water consumption.
○ Residual chemicals in effluents.
○ Condition of joints in pipes.
○ Work practices.
○ Use of safety gadgets.
○ Winding quality of the final dyed yarn.
○ Removal of shade variation and stains if any.

14.6.4 Kier boiling

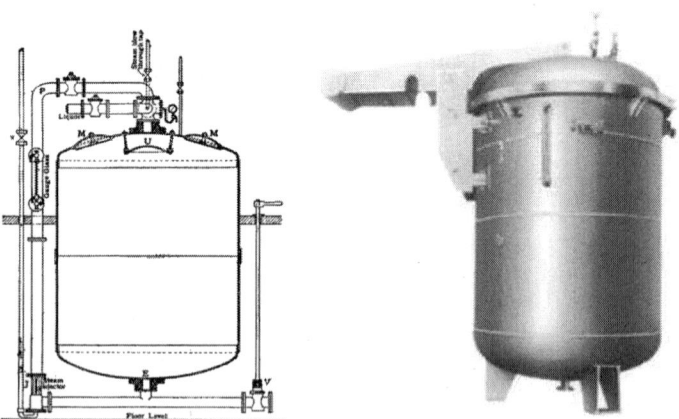

Control points

- Lot size
- Kier boiling parameters like temperature, pressure, time, liquor ratio, caustic to be used, neutraliser, number of washers after kiering
- Persons to be employed

Check points

○ The lot identification.
○ The pH of caustic solution.
○ The pH after neutralizing.
○ The pH of water after washing.
○ Whether the temperature, pressure and time are set as per requirement?
○ Whether the vessel was cleaned thoroughly before starting?
○ Check for leakages of steam and water.
○ Condition of the gaskets and valves.
○ Whether the operators have taken all safety precautions?
○ Condition of the hoist.

14.7   Process – knitting

Knitting process may be of flat knitting type or circular knitting. It may be socks knitting or banian or T-shirt knitting, sweater knitting and so on. Some general points are discussed herewith considering circular knitting.

14.7.1 Circular knitting

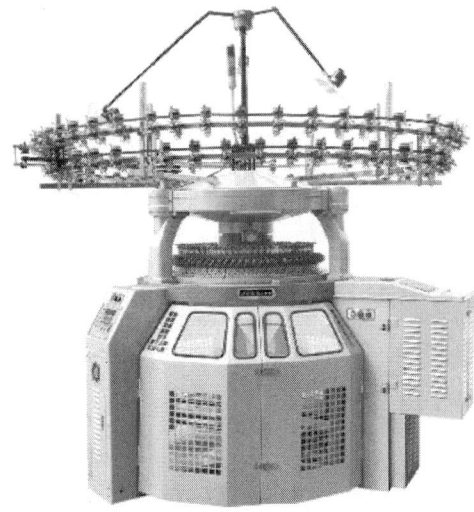

Control points

■ Selecting the yarn count and the yarn parameters to get the required GSM and to suit the gauge of the machine.
■ Selection of knitting parameters.

- Deciding on the speed and production norms.
- Employing of trained personnel and fixing work norms.
- Deciding on the humidity conditions and conditioning of cones before use.

Check points

○ Whether the count of yarn is suitable to the gauge of the machine?
○ Whether the yarns have required strength, elongation, twist, and winding quality is good?
○ Whether the tensions adopted are as per the requirement depending on the knitting design viz. Single jersey, Pique etc?
○ Check for stich cam setting variations.
○ Ensure CONI speed equal and number of coils on CONI equal.
○ If uneven cone runout is found through out the machine, check CONI speed and tension.
○ Check Knot catcher size.
○ Check for choke-up for yarn passage pipe and elbow guide.
○ Whether the conditions of ceramic guides (Eyelets) are good?
○ Whether minimum level of oil is maintained in the tank?
○ Whether the take down of fabric is even from all the sides?
○ Whether the air circulation fan is preventing the fluff accumulation effectively on machine and creel?
○ Whether the air and oil supply pins are working properly?
○ Whether the setting between feeder and needles are as per standards?
○ Whether the eccentricity of cylinder ring and sinker ring are optimum?
○ Whether the speeds of the machines are as per the plan?
○ Whether the required production is being achieved.
○ Whether the quality of fabric in terms of GSM, Tube width and spirality are as per norms?
○ Whether the men employed are adequately trained?
○ Whether the work is allotted as per the norms?
○ Whether the workmen while carrying the works follow the work norms?
○ Whether the humidity and temperatures are maintained as per norms?

14.8   Process – maintenance

14.8.1 Sub process – supervision

Control points

- Planning and procurement of spares, lubricants, tools and gauges of correct quality.

- Use of proper tools and gauges in proper way.
- Handling and preservation of spares, tools, lubricants, gauges etc.
- Planning for procurement of spares and maintaining minimum stock.
- Deciding and maintaining the speeds, temperatures, pressures of the machines.
- Engaging trained and skilled persons for maintenance and operating the machines.
- Educating and training the workmen for correct maintenance and work practices.
- Planning maintenance schedules and implementing.
- Allocating maintenance works as per work norms and monitoring the quality of maintenance.
- Preparing and providing work instructions and norms for maintenance activities.

Check points

- ○ Quality of spares, lubricants, tools and gauges received.
- ○ The condition of the spares removed and the reasons for their wearing out.
- ○ Deviations in machine performance and reasons.
- ○ Alignment of machine parts and setting of various mechanisms and parts.
- ○ Whether there is any part vibrating, which need to be stable?
- ○ Whether the temperature generated by the machines is within control?
- ○ Whether the noise levels are in control?
- ○ Whether the lubricants in running parts are in good condition?
- ○ Whether the time taken for doing the maintenance activities is in control?
- ○ Whether the maintenance activities are planned considering minimum loss of production?
- ○ Whether the activities are carried out as per schedule?
- ○ Whether the men employed are adequately trained?
- ○ Whether the consumption of spares and stores are in control?
- ○ Whether power consumption is in limits?
- ○ Whether the production loss due to breakdowns and maintenance are reducing as per plans?
- ○ Whether the quality of materials produced is as per norms?
- ○ Whether the safety gadgets are functioning as per norms?
- ○ Whether the machine history books and maintenance cards are updated in time?

14.9   Process – garment production

In garment production activities there are number of sub processes like inspection of incoming fabrics, laying and cutting, sewing, surface ornamentation activities like embroidery, sequinsing, bead works, etc., finishing and packing. Let us discuss on the control points and check points of some of the sub processes.

14.9.1   Cutting

Control points

- Techpack of the style to be cut with details of fabric. A sample piece is always better.
- The lay length, number of lays to be laid and total panels to be cut in different sizes.
- Approved marker plan and the patterns.
- Selection of suitable cutting machine.
- Frequency of replacing the cutting blades.

Check points

- Whether the fabric is inspected and approved before issuing for cutting?
- Whether the patterns issued have adequate markings for pattern size, centre back or centre front, fold lines, balance marks, grain lines, construction marks, seam allowances, and are approved by the competent authority?
- Whether the marker plan is approved?
- Whether the number of lays put is as per plan?

○ Whether the fabric is straight without any folds or crease on the lay table?

○ Whether the fabric is adequately relaxed before starting cutting?

○ Whether the cutting machine selected is suitable for the fabric and the purpose?

○ Whether the cutting pitch is maintained by the cutters while cutting?

○ Whether the cutters are using the safety gloves?

○ Whether the safety gloves used is in good condition?

○ Whether the blades are replaced in time and are sharp enough to perform?

○ Whether the numbering and bundling of the panel are as per requirement?

○ Whether the panels are checked before delivering to sewing?

14.9.2 Sewing

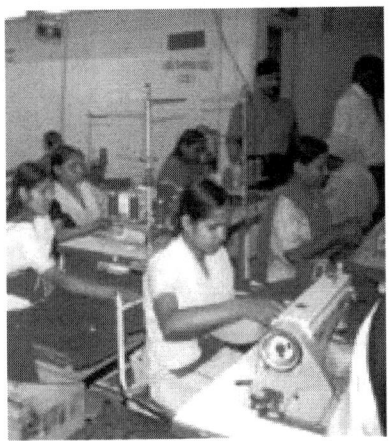

Control points

- Techpack of the style with details of sizes, stitches and seams.
- Number of garments to be produced in each style and size.
- Selection of suitable needle considering the type of fabric.
- Selection of suitable sewing thread to match the fabric and the needle.
- Selection of suitable sewing machine.
- Deciding the number of machines in a batch.
- Selection of tailors depending on the skill requirement.
- Allocation of tailors and machines in a sequence in a batch
- Deciding critical control points for online inspection
- Deciding critical control points for end line inspection

Check points

○ Whether the machine and tailor allocation is as per plan?
○ Whether the bundles received from cutting are as per the production plan?
○ Whether the trims, sewing threads, accessories and buttons etc., received are as per the style requirement?
○ Whether the stitch requirements are displayed at the point of work for the tailors?
○ Whether the work aids provided on the machines are per the style and stitch requirement?
○ Whether the lighting is adequate for stitching and inspection operations?
○ Whether needle guard and eye guards are functional and being used by all tailors?
○ Whether the needles used are correct and replaced as per the predetermined frequency?
○ Whether the broken needles are collected properly using a magnet and deposited?
○ Whether the tailors are getting materials in time for stitching?
○ Whether the online inspectors are clear about the critical points to be checked?
○ Whether the stiches produced are as per the requirement of the style?
○ Whether the required number of garments is being stitched in each hour?
○ Whether the working area is kept clean or not?

14.9.3 Finishing

Control points

■ Techpack of the style with complete dimensions of the final garment, the positioning of trims and labels.
■ Packing details with different sizes to be packed in each carton.
■ Allocation of finishing operators depending on the style requirement and the work knowledge and skill.
■ Washing and ironing parameters.
■ Folding templates.

Check points

○ Whether the garments received for finishing are as per the techpack given and match with the specific purchase order?
○ Whether the number of garments received size-wise match with the

purchase order requirement?
○ Whether the parameters set on washing machine is in line with the customer need and the fabric quality?
○ Whether the temperature and pressure set for steam ironing matches with the fabric quality?
○ Whether the metal detectors are calibrated in time?
○ Whether the garment measurements are as per techpack requirements?
○ Whether the quality of the packing materials is okay?
○ Whether the packers employed are regular employees and their photos are displayed in front of packing entrance?
○ Whether records are maintained to trace the packer for the material and cartons packed?

The above are only some examples, whereas there are still number of processes in textile and apparel industry. Each factory is unique by their technology, size and product range. The customer expectations are changing from segment to segment. Hence, the senior managers in each company should spend their time in preparing control points and check points for their processes, document them and display in the workplace so that the shop floor supervisors can follow that.

15

Normal problems and nonconformities

In this chapter we have discussed some of the normal problems faced in the industry relating to the quality of the materials being produced. Methods of ascertaining and remedial actions also are suggested wherever appropriate. However, this is not the final, as we get new problems when we introduce any new technology or new system. Therefore, the technicians have to always be alert and find the reasons for the defects they get.

15.1 Blow room

- *Low cleaning efficiency* – Lower extraction of wastes than required for that mixing considering the trash content is one of the main reasons for low cleaning efficiency. Increase the wastes if the lint in the wastes is normal or nil. If the beater speeds are lesser than required, we get lower cleaning. Check the beater settings and correct them if needed. Increase the space between the grid bars. Close slightly the air-inlets under the grid bars towards the cotton entry side, and open those on the delivery side. Reduce the fan speed following the beaters by 100 to 200 RPM. If the grip of the feed roller is less, we shall get low cleaning efficiency. Therefore check for the grip. Also check the sharpness of the beaters. Check the synchronisation of the machine working. The blending bale openers should work for 80 to 85% of the time of working of the final machine. If there is a back draught because of not cleaning the wastes under the machines, the cleaning efficiency shall come down.

- *High nep generation and fibre rupture* – The main reasons are blunt beaters, burrs in grid bars, bent pins on beaters, higher beater speed, lower fan speed and excessive feed. A higher beater speed shall give more neps, if the material is not moved out of the beating area effectively. If materials return back to beaters, neps shall generate, hence check the setting of leather flaps, stripper knife etc. Excessive of soft wastes fed, and cottons with more immature fibres are major reasons for neps in opened material. Therefore have a control on the issue of soft wastes to mixing and spread them uniformly throughout the mixing.

- *High variability in the delivered hank* – Improper levels in the hoppers, improper action of feed regulators viz, cone drums, pedals, photocells, direct driving gear motors, etc are the normal reasons for variability in delivered hanks.
- *Formation of cat's tail* – If material movement is less and cottons are over beaten, we get this defect. By sharpening beater edges, increasing fan speeds, increasing the air in-let below the grid bar area of cotton entry, closing the striping knife and beater setting shall avoid cat's tail. The very important step in avoiding cat's tail is to avoid chocking of materials in beaters. Excessive use of cotton-spray oil, water etc., also causes cat's tails.
- *Conical lap* – Conical laps are due to, either higher quantity of cottons coming on one side of the lap, or due to unequal calendar and rack pressures in scutchers. Ensure equal opening of air-inlets under grid bars, replace torn leather lining at the cage, clean the cage thoroughly with emery paper, make pressure on lap spindle uniform on both the sides, remove the pedals and clean thoroughly, and check the pedals where it rests on fulcrum and also pedal fulcrum bar.
- *Lap licking* – Lap licking can be due to excessive addition of soft wastes in mixing, higher rack pressures, lower compacting of laps and excessive dampness in cotton. In case of polyesters, this problem shall be mainly due to static charges and higher bulk of fibres. The problem of lap licking can be reduced by increasing the pressure on calendar rollers, reducing the pressure on racks, increasing the quantity of antistatic, use of roving ends or lap fingers behind the calendar roller nip, blocking of top cage and by reducing the lap length.
- *Patchy lap* – Patchy lap is a result of unopened tufts. Ensure that the mixing is opened thoroughly, and increase opening points if feasible. Check tuft size at the delivery of each beater, and adjust the setting between feed roller and beaters, reduce the gauge between evener roller and inclined lattice, clean the cages, and increase effective suction at cages.
- *Holes in lap* – Holes in the lap can be due to different reasons. Check the cages for damage, and reduce tension draft.
- *Soft laps* – Lower calendar pressure makes the laps soft. Increase the calendar roller pressure.
- *Ragged lap selvedge* – Ragged lap selvedges are mainly due to uneven spots at the edges. Check for the rough spots on the sides of the feed plates, leather linings for the cages, and keep the edges of the scutcher clean.

15.2   Carding

- *Patchy web* – May be due to loading on the cylinder, damaged or pressed wire points, waste accumulation below cylinder under casing or defective under casing.
- *Singles* – May be due to lap licking, less feed in chutes, part of carded web getting sucked by the waste extractor, damaged doffer wire and direct air currents hitting the web.
- *Sagging web* – May be due to insufficient tension draft, very high humidity, worn out key in the calendar roller gears, heavy material fed to card and inadequate calendar roller pressure.
- *High card waste* – High card wastes are due to damaged under casings, higher flat speed, wider front plate setting, closer setting of flats, and higher pressure in suction unit and fibres getting ruptured.
- *Low nep-removal efficiency* – Blunt wire points, too wide setting between feed plate and licker-in, uneven settings, burrs in front plate/back plate, and card wires of coarse type are the main reasons for low nep-removal efficiency.
- *Higher U% of sliver* – Worn out parts, eccentric movement in coiler calendar or table calendar rollers, uneven feed, waste accumulation in material patch, improper settings and loading of fibres on cylinder and flats are some of the reasons for higher U% of card sliver.
- *Bulky sliver* – Slivers become bulky by use of trumpet of a very large size and lower calendar pressure.
- *Higher breaks* – Very small trumpet, worn out trumpet, uneven sliver with bunches of fibres, worn out gears, damaged clothing, air currents disturbing the web, improper temperature and humidity and a very high tension-draft causes higher breakages of web and sliver.

15.3   Combing

- *Inadequate removal of short fibres and neps* – Check head to head and comber to comber noil percent variations, and check the individual heads for web defects, such as uncombed portions due to slippage under feed roller, slippage of fibres under detaching rollers, plucking of fibres by half lap from nipper grip, web disturbance due to air currents due to defects in brush or/in aspirator. Check the machines thoroughly for bent and hooked needles on half lap and top comb, broken needles, nipper grip, feed roller grip, condition of detaching roller cots, condition of the gears driving bottom detaching rollers and damaged air seals in the aspirator box.
- *Short term unevenness* – Prominent piecing waves, drafting waves, uneven fibre control due to worn out top roller cots in draw box,

eccentric rollers in drafting / detaching field, play in draw box drive, high or low tension draft, and improper settings are the main reasons for short term variations in a combed sliver. Check U% and make use of spectrogram diagram to identify the source of the problem.

- *Hank variations* – Single, double or uneven working of sliver on table due to improper selection of tension draft, rough surface of the sliver table, variation in the feeding lap, lap licking while unwinding, etc., are main reasons for variation in hank. If between comber variations are high, check the combers for variations in lap roller feed per nip, draft wheels on draw box, tension drafts at tables, draw box and coiler, and noil level variations.

- *Higher sliver breaks at coiler* – Sliver guides with rough surfaces, coiler calendar rollers having eccentricity or jerky motion, high tension draft, improper balancing of sliver stop motion working on gravity principle, worn out gears, excess parallelisation of fibres in the sliver, improper condensation are the main reasons for sliver breaks. Check whether tension draft between draw box calendar roller and coiler calendar roller is too high causing stretching of sliver, or too low causing slackening of sliver. Check balancing of sliver break stop-motion and ensure that it presses against the sliver very lightly.

- *Frequent coiler tube choke-ups* – If the coiler tubes are loaded with wax and trash, the sliver gets chocked. Clean the coiler tube with a rough rope. If cans are over filled, or the can spring is forcing the sliver to coiler plate, the choke up shall take place.

- *Web breakages at draw box* – Burrs or accumulation of wax/trash particles at trumpet, too much spreading of web, defects in gear wheels, improper tension drafts are the main reasons for breakages in the draw box zone. In cases where the top rollers are buffed badly, the cottons shall stick to top rollers and lap.

- *Breakages at sliver table* – Waxy and rough surfaces of the table, improper tension drafts and piecing waves are the main reasons for breakages on the sliver table.

- *Breakages on comber heads* – A tight or slack web, improper positioning of web trays, unclean web trays, burrs in calendar trumpets, improper calendar trumpet (heavy or light), improper functioning of clearer rollers in detaching section, piecing waves, and the trumpets set too away from the nip of calendar rollers are the main reasons for breaks at comber heads.

- *Detaching roller lapping* – Rough or waxy surfaces on top roller cots, improper functioning of clearer rollers, too wide a setting of web guides are some of the reasons for lapping on detaching rollers. As the detaching top rollers tend to bend at the centre because of loading at both the ends, taper buffing is recommended.

- *Excessive lap licking and splitting* – Improper tension drafts and roller setting, excessive draft in the lap former, uneven lap and tight winding while lap preparation are the main reasons for lap licking and splitting.
- *Note* – Combers are very sensitive to changes in temperature and humidity, and hence it is essential to maintain the required temperature and humidity. In the majority of cases the bad working is attributed to fluctuations in temperature and humidity.

15.4   Draw frames

- *Improper sliver hank* – Check the hank of input slivers and ensure they are as per plan. Check the draft wheels and ensure that the wheels are put to get the required draft. Check the functioning of Auto levellers by sliver test method, (i.e. A %) and ensure that the input voltage is as per norms. Check the pressure on top rollers and ensure them to be as per norms.
- *Uneven sliver* – Check the condition of top and bottom rollers, setting of rollers, the pressure on the top rollers, the condition of the end bushes of the top rollers, worn out or loose wheels, and ensure even feed material. Make use of uster Spectrogram for identifying the source of the problem. Ensure that the slivers do not hit the can surface while getting filled. A bad quality spring in the can make the sliver tilt and spoil the same.
- *Singles* – Stop motion failures are the main reason for singles in draw frame as a can run out is not noticed by the tenter. A very high suction power of pneumafil sucks good fibres and can result in singles. The singles for a short length can also be due to partial lapping on rollers.
- *Cuts in sliver* – Cuts in sliver are mainly due the settings not matching to the fibre staple length. However cuts can also be due to eccentric rollers, worn out end bushes, eccentric coiler shaft drive and grooved calendar rollers.
- *Good fibres in suction waste* – Too close a setting of suction nozzle and a very powerful suction are the reasons for good fibres going in suction waste.
- *Improper coiling* – Non-centring of can and eccentric bottom plate are the main reasons for improper coiling. Also the speed of the can and coilers are to be synchronized to have the required spacing of coils in the can.
- *Higher breakages* – Check the hank and uniformity of sliver, the sliver condensation, condition of gears, the tension draft and ensure smooth surface which comes in contact with sliver/web. Ensure the temperature and humidity to be as per requirement. Check for the

surface of cots, if it is rough, it is likely to lap. Similarly, if the top roller pressure is very high, there shall be lapping on top rollers.

15.5   Speed frames

- *Higher U% of rove* – Inadequate top arm pressures, improper settings, worn out gears or bearings, grooved top rollers, tilted top rollers, wrong selection of condensers, worn out aprons, poor cleaning of draft zone, higher stretch, uneven feed material, sliver splitting in creel, jerks in creel movement, vibrations in the machines are some of the reasons for higher U%. It is always essential to refer the spectrogram and check the spindle and the feed material before taking any action.
- *Higher breakages* – Uneven material, worn out parts, vibrations, insufficient twist, improper draft distribution, fluctuations in temperature and humidity, rough surface in the flyer tube, improper build of bobbins, improper piecing of draw frame sliver, uncontrolled air current etc., are the main reasons for breakages at speed frames.
- *Soft bobbins* – Soft bobbins are due to finer hank, may be due to singles or a finer drawing hank, less number of turns per pressure, and the shift on cone drum (Building mechanism) faster than required, a lower TPI and a lower relative humidity.
- *Lashing in* – Whenever an end is broken and joins to an adjacent end, we get lashing in. Fixing of separators and setting the suction tubes near to the front roller nip shall solve this problem. Moreover we should work towards zero breaks.
- *Hard bobbins* – This is due to a coarser hank; may due to doubles, coarser draw frame hank and lower pressure in top arms. Hard bobbins are also due to higher twist, lesser movement of belt on cone drums, higher turns on the flyer presser, shifted cots in the back zone leading to low pressure, and a higher RH%.
- *Oozed out bobbins* – Malfunctioning of reversing bevels in the builder motion, stopping the machine when the bobbin rails are in extreme positions, and jumping bobbins are the normal reasons for this problem.

15.6   Ring frames

- *Ring cut cops* – Ring cuts are due to the cop diameter becoming more than the limits prescribed by the ring diameter. The possible reasons are the count being coarse, the build not adjusted properly, a bigger ratchet wheel, less number of teeth on ratchet wheel being pushed each time, loose ratchet wheel, vibrating spindles, non alignment of rings in the centre of spindle axis, use of a lighter ring traveller, jammed poker bars, insufficient pressure on top rollers resulting in coarser count etc.

- *Hard twisted yarn* – Hard twisted yarns are due to count becoming very coarse because of inter doubles or lashing in, front roller delivery becoming less due to a loose roller, worn out threads in fluted roller joints, failure of delay drafting mechanism in G5/1 ring frames, loose timer belt driving the front roller in G5/1 ring frames, traverse going out of drafting area, etc.

- *Uneven yarn* – Uneven yarn are due to uneven feed material, improper settings in drafting zone, worn out cots, worn out aprons, improper selection of spacer, low pressure in top arms, eccentric fluted rollers and cots, jammed arbours, vibrating spindles, jammed bobbin holders, improper distribution of drafts between break draft zone and main zone, improper cleaning of drafting zone, lapping in adjacent spindles, traverse going out of drafting area, etc.

- *Soft twisted yarn* – Soft twisted yarns are due to loose tapes, worn out tapes, jammed spindle bolsters, loose bobbin on spindle, jammed jockey pullies, spindle button missing, etc.

- *Higher hairiness* – Higher hairiness are due to worn out rings, worn out travellers, worn out lappet hooks, worn out separators, higher spindle speed, improper selection of traveller, variations in fibre lengths, lower humidity in the working area, too big a balloon, vibrating spindles, etc.

- *Lean built cop* – Lean built cops are due to very fine count, excessive breakages on a particular spindle, high chase length, smaller ratchet wheel, etc. Normally the eccentric spindles and non-centring of lappet hooks are the one who contribute for high breaks and then to lean built cops.

- *Sloughing off at winding* – Loose built soft bobbins, very low chase length, improper combination of winding and binding coils are the normal reasons for sloughing off at winding.

- *Undrafted end* – High twist in rove, lower break draft, low top arm pressure, higher humidity, smaller spacer, channelled aprons and cots are the normal reasons for undrafted ends in spinning.

- *Higher thin places* – Excessive draft, wider roller settings, worn out gear wheels in drafting zone, jerks in working, eccentric movement of cots and fluted rollers, partial lapping on drafting rollers, higher stretch between bobbin holder and drafting zone, broken roving guides are the normal reasons for higher thin places.

- *De-shaped cops* – Not attending to breakages in time, non attending to creel runs out in time and excessive breakages are the main reasons for de-shaped cops.

- *Idle spindles* – The idle spindles add cost to the manufacturing without producing. Some mechanical reason prevents the spindle from being utilised by the sider. The main reasons are non-creeling of bobbins in time, tape and apron breakages, oil flowing to rollers from gears and broken/missing spare parts etc.
- *Slub* – Slub can happen due to improper mixing of fibres, too much variations in fibre lengths, improper opening, low pressure in drafting roller, inadequate drafts applied, lower setting of drafting rollers for the fibre length in use, damages in card wire points resulting in bunches of fibres, lashing in or lapping in spinning, damages in the draft gear wheels and slippage of rollers while drafting.

15.7   Winding

- *Improper splicing* – Normally, spinning mills run different counts on the same winding machine, but shall not be able to match the prism size, the air pressures, timing etc., required for the yarn. The problem is more in coarse counts. Low air pressure, improper mingling chamber and improper setting for opening and splicing are the main reasons for improper splicing.
- *Electronic yarn clearer (EYC) failures* – Fluff accumulation in the measuring slot, low input voltage, blunt cutters, jammed cutters, loose fitting of PCB are the main reasons for EYC failures.
- *Double end* – Failure of EKP (Electronic Knotter Programmer) in cutting the end after a splice/knot, failure of bobbin conveyor belt are the main reasons for double ends found on cones.
- *Stitches* – Variation in cone holder settings, vibrations in cone holder/ drum, and damaged drums are the normal reasons for stitches. The static electricity generated during winding, especially creates stitches in manmade fibre yarn winding, which are prone for static charges.
- *Soft/Bulged cones* – Very low tension, yarn going out of the tension disc, fluff accumulation between tension discs, improper rotation of tension discs are the normal reasons for soft or bulged cones.
- *Sunken nose/base* – Improper fitting of paper cones on cone holder, improper size of paper cones, and improper setting of cone holder are the main reasons for the sunken nose or base.
- *Weight variation between cones* – Variations in tensions between drums; improper setting of either conometer or diameter on cones, malfunctioning of drum sensor and too much variation in yarn count are the normal reasons for weight variation between cones.
- *Shade variation within cones* – A mix up of yarn from different mixing, variation in day to day mixing preparation and addition of soft wastes, ring cut or abrasion of yarn polishing a part of yarn,

very slack tape resulting in a very low TPI, contamination of oil or grease while spinning, exposing the ring cops to smoke or fumes are the normal reasons for shade variation within cones.

15.8   Rotor spinning

- *Neppy and uneven yarn* – Dust accumulation within rotors, mark in rotor grove, damages in rotor covers, lapping on opener roller, damaged wires on opener roller are the main reasons for neppy and uneven yarn in rotor spinning.
- *Stitches* – Stitches on cheeses are mainly due to lapping on the cradle sides, choke up in the traverse path, lapping on drum, damaged cradle bearing, damaged bobbin holder and snap in traverse bar belt.

15.9   Ply-winding

- *Stitches on ply cheeses* – Vibration of cradles, eccentric or damaged tubes, loose adopters on cradle, damaged drum and bent pins in stop motion are the common reasons for stitches in an assembly wound cheese.
- *Singles* – Failures in stop motions like drop pin not falling, contact box pins not acting, defective micro switches, magnet coil not operating, fluff in contact box assembly, carbon depositions etc., are the normal reasons for singles. Entanglement of yarn on drop pins does not allow the pins to fall resulting in singles.
- *Split ends and loose ends in cheeses* – Operators not removing the singles from the cheeses, missing split end preventer, damaged drum, are the main reasons for split ends and loose ends in a cheese.

15.10   Doubling and twisting.

- *Corkscrew* – We get corkscrew when one yarn is straight and other is twisted on it. Mix up of yarns of different counts, and yarns fed with different tensions either in assembly winding or in the twisting machine are the main reasons for a corkscrew. The corkscrew effect is long and continuous when the tensions set for two yarns are different on an assembly winding, whereas short term corkscrew effects are mainly due to improper tensions while joining a broken end in assembly winding. The problem is more where cones are fed directly without making an assembly wound yarn.
- *Snarls* – If one yarn is very loose for a short distance, it results in a snarl. A snarl in the doubled yarn is the weakest spot, and it is very necessary to avoid it. The operator at assembly winder is to be trained

well to ensure that tensions in both yarns are uniform especially while mending a break.

- *Filamentation and drop in strength* – Cut and rough surfaces in yarn path removes the fibres/filaments, from the yarn surface, and reduce the strength.
- *Uneven package length in a T.F.O package* – The variations in the feed package lengths, excessive breakages on a particular drum in a two-for-one twister can lead to uneven package lengths. Ensuring correct weight of packages or using conometers in assembly winding can reduce the variations between packages of TFO yarn. Proper setting of the tensioners can help in reducing the breakages.
- *Zero twisted yarns* – The zero twisted yarns are produced if the magnet in the Two-for-one twister does not hold the supply package and allow it to revolve. Also, the worker might produce a zero twisted yarn while mending a break in twister. In such cases, we can see a zero twist yarn followed by a knot, and the length of the zero twist being equal to the distance from the supply package to the cone being produced on the drum.
- *Multiple fold yarn* – Multiple fold yarns are produced by lashing in of a broken yarn in ply winding to adjacent yarn.
- *Big or improper knot* – This problem is mainly due to improper selection of knot. Depending on the type of yarn and the number of plies the proper knotting method is to be adopted. For fine counts with just two plies, a normal dog knot or a weaver knot can suffice. If the count is coarser, fisherman knot would be more appropriate. Staggered knotting is practiced in case of multi-fold yarns or cabled yarns to avoid big knot at one place.

15.11 Warping

- *Variation in tension within and between beams* – Change in balloon heights as the cones run out, reduction in unwinding diameter of the package as the unwinding progresses are the common reasons for the variation in tension between beams. The difference in the lengths of yarn between respective first thread guide and the head stock, the number of guide points for the yarn in the creel, the angles of deflection in the path of yarns from the axis of the respective packages to the first guide roll in the headstock, and variations in the weight of tension discs and package diameters are the reasons for variation in tension between ends in a beam.
- *Missing ends* – Improper functioning of stop motions, bent drop wires, fluff accumulation in stop motions and entanglement of broken yarn with drop pin are the main reasons for missing ends.

- *Crossed ends* – Broken ends not drawn correctly and improper insertion of lease rods are the reasons for crossed ends.
- *Poor quality of beam preparation* – The build of the beam should be firm, which can be achieved by uniform tensioning with required weights. The surface of the yarn building on beam should be even and free from ridges from one selvedge to another. For this we need to ensure uniform spacing of comb dents, matching the comb width with the beam width, minimum variation in tension within and between ends, matching of distance between beam flanges with the length of the drum and concentricity of beam flanges. The yarn sheet from the beam at subsequent processes should unwind smoothly without entanglements and breaks. We need to ensure effectiveness of stop motion and brakes, proper mending of end breaks without introducing kinks snarls or cross ends, good condition of beam flanges and proper working of creel fans. The yarns in the beams should not have any frictional damages, for which we need to ensure smooth surface of the drum, absence of cuts in parts of the machine coming in contact with yarns, and suitability of warping speed for the type of yarn and the count. All the warper's beams in a set, or all sections in sectional warping should have same length of warp sheet, for this the length measuring system should be accurate. During unwinding of beams, there should be minimum variation in tension from start to finish, for this the prescribed minimum diameter of the barrel is to be maintained.

15.12 Sizing

- *High size pick up* – Excessive pressure on squeezing rollers, slow speed in sizing allowing the yarn to remain for a longer time in sow box and use of more wetting agent than required are the main reasons for high size pick up. A higher size pick up can make the yarn brittle and rough, and also shall add to the cost of manufacturing.
- *Low size pick up* – Lower pressure on the squeezing rollers, a higher twist in the yarn, improper preparation of size recipe, lower temperature in sow box, higher viscosity of the size and insufficient time given for yarn to stay in sow box are the main reasons for low pick up. A low pick up results in lower strength of yarn and hence can lead to breaks. In some instances, the weavers want more weight of fabric, and for that they insist on higher pick up, and the sizer adds weighting agents. This is not a good practice, as it reduces the life of loom parts. It is essential to work out the required count of yarn to get the prescribed weight per square meter of fabric, but the weavers go for finer count to get more length of fabric and end up

with low weight per square metre of fabric.

- *Low elongation at break* – This is due to excessive stretching of warp during sizing. Some sizers purposefully keep a higher stretch to get an advantage of higher length of fabric, but this leads to higher breakages at loom, and the loom efficiency drops down, and there shall be a big loss. High stretch also results in higher lappers and cross ends. It is essential to measure the stretch and maintain zero stretch to get a higher elongation at break, a good working of loom and a fabric with a good feel.

- *High moisture content* – Low temperature of drying cylinders due to low pressure and condensation not removed periodically are the main reasons for high moisture content resulting in sticky ends, soft beam and higher breaks.

- *Low moisture content* – Excessive drying results in low moisture content, which leads to brittle yarn and high breaks.

- *Excessive lappers* – These are due to stickiness and high breaks while sizing. This leads to more migration of ends and cross ends, leading to higher breaks.

- *Uneven size pick up* – Improper size preparation with undissolved parts, improper addition of size into sow box, allowing the steam to condense in sow box diluting the size, improper surface of rubber roller in sow box, uneven pressure on squeezing rollers, frequent stoppages of the sizing machine and jerky motion are the normal reasons for uneven size pickup.

15.13 Weaving

- *Warp streaks* – Warp streaks are narrow, barre and dense stripes running along the warp direction. Main reasons are the variation in density of adjacent group of warp ends due to non-uniform dent spacing, wrong drawing-in, or count variations. Also, the variations in lustre, reflectance of dye pick-up of adjacent groups arising out of differences in raw materials, blend composition or yarn constructions contribute for streaks.

- *Reediness* – These are very fine cracks or lines between groups of warp threads, caused due to excessive warp tension, late shedding, use of coarse reed with more number of ends per dent, bent reed wires, improper spacing of reed wires, wrong drawing, and insufficient troughing of shed, i.e. tension difference between top and bottom shed lines during beat up.

- *Weft bar* – It is a band running weft-wise across the full width of the cloth. The normal reasons are the periodic medium to long term irregularity in the weft yarn, count difference in weft, excessive tension in the weft feed package, especially in filaments, variability in pick density and difference in twist, colour or shade of adjacent group of picks, difference in blend composition or in the cottons used.

- *Weft crack* – It is a thin place or missing weft across the body of the fabric. The main causes are improper setting of anti crack motion, loose fitting of reed, loose or worn out crank, worn out crank arm, worn out crank shaft bearings, loose belt, worn out duck bills and beaters, weft fork not functioning properly, faulty take up, brake motion not acting instantaneously, shuttle striking on the weft fork due to weak picking, swing rail worn out, weaver not adjusting the fell of cloth properly at the time of starting a loom, and gripper not holding the weft firmly.

- *Thick and thin places* – These are similar to weft bar, but unlike weft bars, it repeats at intervals. They are mainly due to irregular let-off, incorrect setting of holding and releasing pawls on the ratchet wheel of take-up motion, gears of take-up motion not meshing properly, and gear wheel teeth worn out or broken.

- *Weft loops* – Loops project from the surface of cloth either on one or both sides of a cloth because of a small portion of weft getting caught by the warp threads. The main reasons are late shedding, low warp tension and use of bad temples.

- *Box marks* – Box marks are due to something bruising or staining the weft while it is in or near the box. Main causes are dirty boxes, shuttle riding over the weft, oil from shuttle tongue, dirty shuttles, weft flying about too freely, oil splashes from loose cranks, oily spindles and buffers and dirty picking stick for under pick.

- *High incidence of warp breaks* – Excessive warp tension, blunt or loose shuttle tip, rough shuttles, too small or too big shed formation, bottom shed line beating down on slay race, jerky movement of healds, too early or too late shedding, race board badly worn out, healds catching each other, sharp or rigid reed wires, warp size accumulation on reed, pirns projecting above or below shuttle, improper sizing, improper humidity in the loom shed, a weaker warp yarn, a higher speed of loom, more number of ends per inch for the count being used, less air space in reed are the main causes for excessive warp breaks.

- *Weft breaks* – High weft tension, improper build of pirn, knots at the nose or chase of pirns, back stitches in cones fed as weft in shuttle-less looms, rough and damaged surface of pirns, shuttle tongue not in level, rough places inside the shuttle, damaged nylon loops, sloughing off or loosely built weft package, shuttle eye chipped or broken, weft trapped in the box, selvedge ends cutting the weft, weft fork too far through the grate, rough box fronts or shuttle guides, improper alignment of cone in weft feeder, lower twist in weft resulting in weft opening out in air-jet looms, grippers missing the picks, improper knotting of tail ends, and rough handling of cones are the main reasons for higher weft breaks.

- *Shuttle traps* – Entangled warp ends due to fluff falling on the warp, broken warp end entangled to adjacent end, knot with a long tail resulting in entanglement, snarls in yarn getting entangled, too much hairiness in yarns, weak picking, faulty shuttle checking, gear wheels slipping due to broken teeth, loose stop rod finger, and uneven joint of flat belt are the normal reasons for shuttle trap.

- *Shuttle flying* – Fibrous yarns, knots with long tail ends, slack warp, uneven race board, small sheds, bottom line too high, worn pickers, swells giving twist to the shuttle as it leaves the box, early picking, late shedding, unbalanced shuttle, box spindle not set properly, box front not set properly and missing shuttle guard are the main reasons for shuttle flying.

- *Smashes* – Daggers not working, frog spring ineffective, bad shuttle, improper boxing of shuttle, worn out picker, worn out transfer hammer, damaged pirn and entanglements are main causes of smashes.

- *Bad selvedge* – Improper shuttle wire tension, bent shuttle jaw, shuttle crack, more tension on selvedge yarns, late shedding resulting in rubbing of shuttle to the selvedge and improper selection of selvedge weave for the fabric being woven are the main reasons for bad selvedge.

- *Broken picks* – A filling yarn that is broken in the weaving of a fabric appears as a defect. Improper functioning of weft stop motion results in broken picks undetected and going in to the fabric.

- *Bullet* – Bullets are low twisted double yarn seen weft wise in fabrics. Those are generally zero twisted parallel yarns. Practical causes of faults are improper functioning of bunch motion, incorrect yarn path through spindle, loose tensioners, capsule and spring working, insufficient yarn as bunch and knot is not applied after removing bunch yarn

- *Half pick* – In case of rapier looms, if the second rapier does not collect the weft, it shall stop in between, and we get half pick.

- *Broken end* – A defect in fabric caused by a warp yarn that was broken during weaving or finishing.

- *Coarse end* – Warp yarn that has a diameter too large, too irregular or that contains too much foreign material to make an even, smooth fabric.

- *Coarse pick* – Filling yarn that is too large and imperfect to appear to advantage in the final cloth.
- *Slough off* – Weft yarn has slipped from the pirn. Proper monitoring of strength and chase in pirn winding can solve this problem.
- *Thick end and thick picks* – Higher diameter in yarn for a short distance can be due to improper piecing at spinning preparatory or drop in pressure on the drafting rollers for a short time. This also can happen due to not removing of spinners double, not piecing the end properly by removing the lapped materials, accumulation of fluff in condensers, cradles and in the necks of the top rollers.
- *Double end* – Two ends that weave as one. This happens because of migration of a broken end to the adjacent reed space along with the neighbouring end
- *End out* – A warp yarn that was broken or missing during weaving.
- *Fine end* – A defect in silk warp yarn consisting of thin places that occur when some of the filaments that should be in the warp yarn are absent, generally caused by improper reeling. Warp end of abnormally small diameter, i.e. long thin places of class I1 and I2 also is referred as fine end.
- *Jerk-in* – An extra piece of filling yarn jerked by the shuttle into the fabric along with a regular pick of filling.
- *Knot* – Knot is defined as a knob or lump formed by interlacing portions of one or more flexible strands or a quantity of yarn, or thread, which varies with the fibre; it consists of a set of coils. Control in pirn winding, the winding to binding coils ratio can solve this problem.
- *Loom bar* – A change in shade across the width of a fabric, resulting from a build up of tension in the shuttle before a filling change.
- *Loom barre'* – Repetitive selvedge-to-selvedge unevenness in woven fabric usually attributed to a mechanical defect in the let-off or the take-up motion.
- *Misdraw (Colour)* – In woven fabrics the drawing of coloured yarns through the loom harness contrary to the colour pattern and/or design weave is termed as misdraw. In case of warp knits misdraw is the drawing of coloured yarns through the guide bars contrary to the pattern design.
- *Mispick* – A defect in woven fabric caused by a missing or out-of-sequence yarn.
- *Reed mark* – A crack between groups of warp ends, either continuous or at intervals, which can happen due to damaged reed or improper spacing of dents.
- *Reed streak* – A warp wise defect attributable to a bad reed like uneven reed space, bent reed wire, slant wire, damaged reed wire etc.

- *Set mark* – Defect in woven fabric resulting from prolonged loom stoppage. Because of the humid weather and the fine dust present in the atmosphere, the cloth exposed shall get slightly different colour and also some relaxation takes place. A combined effect gives a line in weft direction.
- *Shade bar* – A distinct shade change of short duration across the width of the fabric. This is normally due to a mix up of weft with different property.
- *Stop mark* – Narrow band of different weave density, across the width of a woven fabric, caused by improper warp tension adjustment after a loom stop. A well trained weaver can reduce this type of defects.
- *Tight end* – Warp yarn in a woven fabric that was under excessive tension during weaving or shrank more than the normal amount.
- *Pilling* – Fibre filaments that break in yarn due to friction leaving small clumps of loose fibres on the surface
- *Float* – Slack warp and Faulty Pattern Card are the main reasons for a float in a woven fabric.
- *Pin marks* – Poorly adjusted temple pins or damaged pins can lead to pin marks.
- *Contamination of fluff* – Different fibres or foreign materials get mixed during spinning, winding or in weaving preparation stage, causing visual objection in fabric. The causes are improper cleanliness, not properly cleaning the machines after each doff and lot changes, improper suction of drafting zones of gill boxes and roving, improper cleaning of scrapper and scrapper plate after every lot change of doff, not using of curtains for partition of machines running on different colours, overhead cleaners of ply winding and ring frames blowing dust on running spindles or drums, material not covered to avoid fly and fluff accumulation, use of compressed air for cleaning machines while in working or while adjacent machine is working and use of common return air ducts and running different coloured fibres in the shed.

15.14 Soft package winding of yarn for wet processing

- *Too soft yarn package* – A lower tension on the yarn is the reason for soft packages, which can be due to a lower tension weight applied, yarn passing out of tension guide, and lower speed of winding.
- *Too hard yarn package* – This is mainly due to application of higher tension on yarn while winding.
- *Side stitches* – Improper traverse of the yarn, vibrations in winding

head, and a loose adopter are the normal reasons for getting stitches in the yarn package.

- *Ribbon formation* – Malfunctioning of ribbon breaker and improper cleaning of drums are the reasons for Ribbon formation.
- *Count mix-up* – Improper colour codification and lack of education to the workmen and staff are the main reasons for mix up of cones.

15.15 Scouring

- *Excessive fluidity value* – Oxycellulose formations during scouring and excessive bleaching due to use of sodium hypochlorite are the reasons for excessive fluidity value.
- *Kitties even after processing* – Improper swelling of kitties due to inadequate caustic in the liquor.
- *Tendering of yarn* – Formation of oxycellulose due to too high concentration of scouring chemicals.

15.16 Yarn dyeing in package form

- *Shade variation within cheeses* – Variations in package density within the package, mix-up of yarns of different count/TPI/cotton within the package, package touching each other during processing are the normal reasons for shade variations within package.
- *Shade variation between cheeses* – Variation in package density between cheeses and improper draining of liquor are the main reasons for variation in shade between cheeses.
- *Uneven dyeing* – Improper preparation of dyeing recipe, inadequate wetting, inadequate steam pressure, inadequate temperature of dyeing, improper additions of chemicals, and inadequate liquor ratio lead to uneven dyeing.
- *Stains* – Stains of rust and oil can come from the hoist while transporting yarn packages. Stains can also come from the drier and can be controlled by proper cleaning of drier and drier filter. The stains can also come due to other dyestuff particles on wet cheeses, which can be controlled by covering the empty cheeses with clean paper, cloth or a polythene sheet.
- *Improper shade or tonal difference between lots* – Improper weighment of colours and chemicals, improper dissolution of colours and chemicals, lack of control in the time of treatment and increase in temperature and pressures, improper matching of samples and shade, change in some of the colours or chemicals and change in cotton lots are the main reasons for getting shade difference between lots.

- *Poor fastness properties* – Improper washing resulting in non-removal of unfixed dye-particles, inadequate temperature and pressure during dyeing, variation in process parameters are the normal reasons for poor fastness properties. A wrong selection of dyestuff can also cause poor fastness properties.
- *Brittle yarn* – Over drying of dyed yarn and inadequate softening agents while finishing are the main reasons for brittleness in the yarns. Waxing the yarns after dyeing shall reduce the brittleness of the yarns and reduces its friction value.

15.17 Fabric dyeing

- *Patchy dyeing* – Improper scouring and bleaching, uneven surface of padding rollers, and folds while padding is the reasons for patchy dyeing apart from improper dissolution of dyestuffs.
- *Side to side variation in tone* – Variation in temperature across the machines, addition of chemicals at one side only and improper batch tensions are the normal reasons of variation in shade from side to side.
- *Bad selvedge* – High temperature, pre-dissolution of chemicals, slippage of goods, longer yardage for vat dyeing on jigger, selvedge remaining exposed for paf batch dyed goods and uneven fabric widths are the main reasons for bad selvedge.
- *Tailing effect* – Loading of long batches in dye bath, high substantive dyes for padding and level difference in padding trough are normal reasons for this defect.
- *Streakiness* – Uneven absorption of dyestuff and particularly when dyeing fabric made of high twisted yarns and variation in yarn parameters are the normal reasons for streakiness
- *Dye mark* – Uneven dyeing of the fabric may be due to improper wetting, unexpected stagnation while dyeing.

15.18 Finishing

- *Precipitation in finishing bath* – Incompatibility of finishing chemicals, use of anionic wetting agents/soaps with cationic softness agents are the normal reasons.
- *Uneven finish* – Inadequate cooking, mixing and feeding of finishing ingredients are the normal reasons for this type of problems.
- *Uneven optical brighten blueing* – Variation in time and temperature, inadequate mixing of tinting colour and use of resins along with optical brightening agent are the normal reasons for this defect.

- *Yellowing* – Prolonged exposure of the material at elevated temperature due to over stoppage of the machine is the main reason for yellowing.

15.19 Mercerization

- *Uneven mercerisation* – Improper tensioning, improper preparation of caustic solution, variation in time and temperature of treatment are the main causes of uneven mercerisation. Improper pressure in squeezing rollers after padding, variation in hardness of the padding bowls, variation in the diameters of the padding bowls are the main reasons for variation in mercerisation in fabric mercerisation.
- *Uneven shade after mercerisation* – Uneven singing is one of the main reasons for uneven shade after mercerisation. This can happen due to improper adjustment of flame height because of variation in the voltage in case of electrical singing, and variation in the gas pressure in case of gassing.
- *Excessive consumption of caustic soda* – Excessive pick up of caustic soda at the padding stage, inefficient washing, leakage in underground storage tank and pipe lines, and not having enough capacity to store caustic soda solution are the main causes for excess consumption of caustic soda.

15.20 Roller printing

- *Scratches* – Gritty particles in colour paste cutting the surface of the roller deeply enough to show them when printed.
- *Snappers* – Loose threads from the cloth escaping under the lint doctor, bits of the dried up paste and other hard particles cause snappers, which are large double stripes of colour running along the length of the cloth. Normally two stripes of colour with a white centre are seen.
- *Lifts* – These are also called as minute snappers, occurring usually at regular intervals, which are normally caused by some hard particles like a metal piece getting embedded in the engraving of the roller and protruding from it, thereby lifting the Doctor blade temporarily.
- *Streaks* – Fine lines parallel to the selvedge may be caused due to scratches on the engraved roller or a cut on the colour doctor edge because of some grit in the colour paste.
- *Scumming* – Soiling of the cloth by one or more colours due to insufficient scrapping of the print paste from the engraved portion. It may be due to rough doctor blade, improper adjustment of doctor

blade, badly faced roller or defective printing paste preparation.

- *Scrimp* – Creases remaining in the cloth while printing give rise to the defect of non-printing underneath the fold.
- *Uneven printing* – Uneven pressure due to faulty lapping or improper feed of the print paste, too much polishing of certain parts to eliminate scratches, give uneven impressions.
- *Lobbying* – This is an uneven printing due to slippage of roller round its mandrel, due to improper fitting.
- *Back-grey crease* – Improper feeding of the fresh back grey or due to heavy soiling, inefficient washing and improper drying of used back-grey are the main reasons for this defect.
- *Back grey stitch impressions* – Although this is an inevitable defect because of the stitch impressions, it can be minimised by using the suitable sewing thread.

15.21 Screen printing

- *Choking of screens* – High viscosity of printing paste, improper profile of squeeze blades, improper cleaning of screens, deposition of thickening agent under or over the screens and frequent stoppages of printing are the normal reasons for choking of screens.
- *Misfitting of the design* – Improper tension of screens, worn out thermoplastic coating, deviations in blanket guide controlling system, loose end rings, and pressure roll not working, insufficient quantity of colour in the screen, defective working of printing head, magnetic clamps and inadequate temperature are the normal reasons for misfit of the design.
- *Stains* – Stains on the garment can be caused by a variety of factors. The printer could get a little over zealous about his inking or the folders could have a Java disaster or the mill could leak a bit of machine oil during the sewing process. Stains are clear defects and the printer should be informed about even subtle discolouration on the garment. The solutions include good work practices, wiping the machine and floor thoroughly after oiling, ensuring that workers keep their hands clean, using of dry lubricants wherever feasible, keeping the work area always clean and covering the materials with clean covers.
- *Conveyor stain* – Improper drying, improper cleaning of conveyor, improper speed synchronisation between the machines and the dryer, uncleaned nozzles and strainers are the normal reasons for this defect.
- *Blanket stain* – Failure of water supply or the washer pump, uneven thermoplastic coating or lines on thermoplastic are the normal reasons for this defect.

- *Misprint or no print on selvedge* – Improper setting, defective guiders, and uneven width of the fabric at stitches are the reasons for misprint on selvedge.
- *Design not washed out properly* – Positive permeable to light rays, too warm a drying before exposure, insufficient contact pressure, too long a delay before exposure, copying emulsion too cold and exposure time too long are the reasons for design not getting washed out properly.
- *Slippage on the cloth* – Frames not properly roughened, adhesive not evenly applied causing bubbles on the surface and cloth strip not applied properly to avoid water or colour penetration.
- *Pinholes* – They are tiny breaks in the emulsion that coats the screen and appear as small dots of ink where there ought not to be any. They can be removed, (except in garment dyed shirts), with a spotting gun. Unfiltered photo emulsion in use, dust in the working area, insufficient light source and low concentration of hardener are the normal reasons for pinholes. Verify the screen thoroughly before taking it for printing.
- *Pilling of the lacquer* – Too thick an emulsion coating, improper degreasing and wrong proportion of hardener are the normal reasons for this defect.
- *Placement* – There are general rules for placement of an image on a garment, but since all garments are of varying dimension and proportion, exact placement can be a judgment call. It also depends on the size and shape of the image itself. The rules of thumb are; full front- 3-4 inches down from the collar, full Back- 4-6 inches down from the collar and left chest-bottom aligned with bottom seam of sleeve. All of these are general rules, however, in the end the decision on the part of the printer considering aesthetic look is important. If there is an intended placement that deviates significantly from the above guidelines then one should make it clear to the customer before printing. A normal practice is taking a photocopy of the image at full size and sticking it onto the shirt to see how it looks. If you determine that it needs to have an unusual placement then send your 'mock-up' to the printer.
- *Consistency of placement* – Minor deviations are found in the placement from shirt to shirt. The printer generally loads the shirt onto the platen the same way every time but shirts can be quite irregular dimensionally. Hence often the printer must make a judgment. If you have exceedingly specific criteria for placement consistency you should make this clear from the outset.
- *Colour correctness* – Because the gamut for process printing on garments is much smaller than most other printing methods colours

do not match exactly. A carefully engineered separation and a skilled inker should be able to deliver a pleasing print that captures the essence of the range of tones and the levels of contrast in the original. Often touch plates are used to achieve colours that are out of range. Process printing on dyed shirts yields a much narrower range than on white shirts. For spot printing the range of colours is similar to offset. Specifying colours from any of the standard matching systems, including pantone, focoltone and trumatch shall help the printer. It also helps if their ink department and print areas have a good graphic standard 5000k light source to match colours.

- *Colour consistency* – Maintaining colour consistency in halftone printing is a challenge. Hues in process are determined by the proportional densities of the 4 process colours. These proportions can be disrupted by many factors that determine the amount of ink flowing through a particular screen. The most common cause is uneven level of platens which changes the critical off contact distance, often causing a visible shift in the hues. Process colours are difficult to match from run to run if all of the critical variables are not recorded and controlled. The tools necessary to control this myriad of interrelated parameters are not standard equipment in the vast majority of screen print shops. The absence of tools such as deltascopes, colorimeters, off contact gauges, and print pressure meters will indicate to the prospective shirt buyer that the shop may not be capable of the consistency. The solution include colour matching from run to run by employing a structured colour matching system, presence of a quality digital scale and a catalogue of achievable ink colours.
- *Colour smear* – In printing the colour gets smeared by distorted patterns. Proper colour paste, applying required pressure while printing and avoiding lateral movement of screens while placing on fabric for printing or removing after printing can prevent this problem.
- *Dye migration* – This is an effect generally seen on shirts containing polyester. Since the dyes used for garments don't readily bind themselves to polyester fibres the colour can affect the printed area. This effect can be seen immediately after curing or can appear weeks after. Red shirts with white ink are the most notorious for this effect but many other combinations can also give trouble. Selection of dyes compatible with polyester and strictly adhering to process parameters and timings is very important.
- *Scorching* – Scorching is caused by improper heating of the shirt between colours on press in the flashing stage or in the main dryer during curing. Scorching can evidence itself in a range of hues, from almost undetectable yellow to a Cajun blackened. Plastisol inks are the most durable but require heat to cure. Large areas of yellow or

brown as well as brittle fibres are indications of a scorched shirt. A delicate balance of temperature and time to properly gel or cure the inks is to be made and if diligent measurements are not taken shirts can be easily torched. Occasionally, this phenomenon can be caused by sizing left in the shirts from the mill. Under normal curing conditions this sizing can create a light, yellowish cast.

- *Improper curing* – Improper curing can be seen as inks loose much of their vibrancy or opacity after washing. This should not be confused with fibrillation. One of the most carefully monitored factors in screen printing with plastsols is the curing process. The ink must reach a certain temperature to completely cure.

- *Fibrillation or frosting* – This effect occurs on light shirts and is often confused with improper curing. The effect is visible on prints employing transparent inks using the whiteness of the shirt to achieve certain bright hues. When these inks are washed the lack of a heavy plastic coating allows some of the unprinted fibres to break through the ink layer and dull out or "frost" the image. This has recently become more of an issue as market is demanding heavier weight shirts that feel smoother. The fibres in these super heavyweight garments are the most susceptible to this effect. Process printing is vulnerable to frosting as all of the inks used, except black, are transparent. Adhering strictly to the process as designed while developing the sample, and training the people adequately is very important to avoid this problem.

- *Distortion* – The flexible nature of fabric can yield a distorted image if not loaded correctly. The adhesive that is used to hold the panel on the platen can catch part of the garment when it is being loaded and pull it out of shape. There are loading techniques that can alleviate this effect but certain shaped prints, such as hard geometric boxes, will show distortion much more than others. Training the operators adequately is the solution for this problem.

- *Opacity* – There is no specific benchmark for opacity. In halftone printing it is especially problematic to balance dot gain and opacity considerations. On light shirts one should not be able to see the weave pattern of the shirt thorough the ink, even under minor stretching. On dark shirts the problem is compounded by the need to cover the shirt colour with a thick enough layer of opaque lighter colours without making the shirt stiff. In most cases the level of acceptability is a judgment and one should know poor coverage when seen. The solutions include training the operators adequately and educating the customers on basic concepts can reduce the grumbling.

- *Poor wash fastness* – Improper curing of ink leads to poor wash fastness. Adhering strictly to the process as designed while developing the sample, and training the people adequately is essential to overcome this problem.

- *Registration* – The registration tolerances of the various presses used by screen printers range wildly. Any gap between colours that are visible from more than a foot or two away are generally not accepted. A well trained operator with decent well tuned equipment should be able to make product with very little or no visible error. The best way to achieve a pleasing graphic image is to butt register the separations, which requires nearly-perfect registration to print successfully.

- *Hand* – This term describes the amount of ink on a shirt. In certain printing styles, such as athletic, a heavy deposit is acceptable and even, to a degree, expected. In most other styles of printing any large ink area that stiffens the fabric is objectionable. In extreme cases the weight of the ink can be felt and the print will not breathe, causing a nasty adhesive effect on the wearer's chest on summer days. Developing a library of techniques to achieve decent coverage is suggested.

- *Colour out* – While printing, if the colour paste runs low in the reservoir resulting in blank skips in the print pattern it is called as Colour out. Continuous monitoring of the level of the colour pastes can overcome this problem.

- *Scrimps* – Scrimp is a printing defect characterized by lengthwise strips of fabric that are unprinted. This can happen because of the folding of fabrics length wise and not getting spread properly on the printing table.

15.22   Circular knitting

- *Barre* – Barre is an unintentional repetitive visual pattern of continuous bars and stripes usually parallel to the filling of woven fabric or to the courses of circular knitted fabric. Mixing of yarns of different lots or counts, fibre micronaire variation, fibre colour variation, yarn linear density variation, yarn twist variation, yarn hairiness variation, knitting tension variation, improper mixing of cotton from different origin, improper mixing of cotton from different varieties and improper mixing of cotton grown in different seasons, difference in tension between the feed packages, yarns/filaments of different elongation properties fed without proper control of tensions, improper stitching cam setting, uneven take down pull, over stretching of fabrics and exposing of cones to sunlight of a long time are the main causes of barre attributed to knitting operation. Barre can be caused by physical, optical or dye differences in the yarns, geometric differences in the fabric structure or by any combination of these differences. It is basically a visual phenomenon and any property of yarn which makes it 'look' different from the adjacent yarn in a fabric would result in this defect. All barre is the consequence of subtle differences in yarn reflectance between individual yarns in the knit structure. Any mechanism that can change the reflectance of a yarn in a knit structure is a potential barre source A barre streak can be one course or end wide or it can be several - a "shadow band". It is not the inadequacy of the raw material property which results in barre; it is the inconsistency or the variability of the particular property which results in Barre.

- *Holes in fabric* – A hole or a press off is the result of a broken yarn at a specific needle feed so that knitting cannot occur. Fluff accumulation on the knitting creel if fall on the needle zone, or the knots and slubs if present in the yarn are the main reasons for holes in the fabric. Yarn imperfections in which one or several yarns are sufficiently damaged to create an opening also contribute to this problem. A good and clean working area is very essential. The blowing or suction fans can be provided to prevent fluff accumulation on the cones, yarn and the knitting area.
- *Lower GSM* – Count of the yarn finer than the required, a lower loop length, and lesser course and wales are the main reasons for lower GSM of knitted fabric.
- *Spirality* – A higher twist in the grey yarns, improper steaming or conditioning of yarn, take-up roller set in a slanted position are the main reasons for Spirality.
- *Loop to loop variation* – Variation in loop parameters at micro level such as loop-to-loop variation in dimension, geometrical shape of loop and localized variation in loop density can be reduced by improvement of knit structure by better understanding of mechanics of loop formation, fluidity of knit structures and their influence on quality of knit fabrics. The quality of hosiery yarn viz. the uniformity and twist factor for the staple length of fibres used has to be considered with due weightage to these aspects. The machine parameters such as gauge, needle type, cam type, yarn feeding system, number of feeders, take down system, cloth rolling or spreading are to be monitored and controlled.
- *Tight loops* – This may take the form of a shadow (several courses involved) or a discreet line (one course involved). It will normally show up as a dark or dense line or shadow
- *Slack loop* – Slack loop shows up as a sheer or light line
- *Snagging in ladies suits* – A snag is a desirable surface loop of varying size on woven or knitted fabrics often caused by catching on sharp

points or objects. To prevent this ensure adequate twist in the yarn, firm fabric construction, proper handling of materials, and smooth surface at all fabric moving and storing areas.

- *Bird's eye* – In knitting, an unintentional tuck stitch is called as a Bird's Eye. The tension variation and hairiness are the main reasons for this.
- *Dropped stitch* – Defect in knitted fabric; recurrent openings in one or more wales of a length of knitted fabric because the stitches did not knit.
- *Needle line* – Lengthwise streak in knitted goods caused by improper alignment of a needle or its incorrect spacing cause the Needle Lines. This is caused by needle movement due to a tight fit in its slot or a defective sinker.
- *Bowing* – A line or a design may curve across the fabric. This bowing is the distortion caused by faulty take-up mechanism on the knitting machine.
- *Streak or stop mark* – A straight horizontal streak or stop mark in the knitted fabric is due to the difference in tension in the yarns caused by the machine being stopped and then restarted.
- *Skewing* – Skewing effect is seen as a line or design running at a slight angle across the cloth.
- *Slub* – Unexpected sudden thickness in the yarn
- *Press-off* – Condition in knitting when a yarn breaks or fails to knit
- *Run* – In reference to knit goods, a series of dropped stitches is called as a Run, as a run in hosiery.
- *Scrimps* – A wrinkle caused by excessive strain, tension, or pressure on a fabric.
- *Boardy* – The knitted fabric becomes boardy (a stiff or harsh hand) when the stitches have been knit very tightly.
- *Cockled or puckered* – Uneven stitches or yarn sizer result in cockling or puckering.
- *Dropped stitch* – This is an un-knitted stitch caused either by the yarn carrier not having been set properly or the stitch having been knitted too loosely.
- *Run or ladder* – A run or ladder indicates a row of dropped stitches in the wale.
- *Tucking* – This is the result of an unintentional tucking in the knitted fabric. This is also called the bird's eye defect.
- *Float* – This is caused by a miss stitch which is the result of failure of one or more needles to have been raised to catch the yarn.
- *Ladder* – Wales collapse in straight line
- *Missing plush loop* – Malfunctioning of loop

- *Stains* – Stains are mainly due to improper house keeping, oil leakages in the machines, improper handling, uncleaned containers of materials and lack of awareness among the employees and the management.
- *Thin yarn* – Variations in count, singles in yarns, sliver or rove, partial lapping in speed frame drafting rollers, are normal reasons for a thin place in the yarn. They are classified as H and I faults as per classimat.
- *Thick yarn* – Variations in counts, doubles in sliver and rove, lashing in, low pressure on drafting rollers are normal causes for thick yarn. These are classified as E, F and G faults in classimat.

15.23　Embroidery defects in garment manufacturing

- *Bunching at corners* – Where the corners of lettering or shapes are not sharp and crisp but are bunched up or distorted. Usually this is caused by too much thread in the corners due to poor digitizing. This includes not using appropriate stitch selection, not using "Short" stitches in corner, and poor stitch balance i.e. thread too loose. It can be corrected by digitizing properly by using appropriate stitch selection, and "short" stitch cornering, and correcting stitch balance.
- *Embroidery too thick* – Too thick and uncomfortable embroidery can be caused by too high stitch density or not using the correct backing for the application. It can be corrected by digitizing properly by using appropriate stitch selection, using fewer stitches, and using "short" stitches on corners, balancing the stitch, using smaller thread size and the correct backing (correct type and weight).
- *Fabric damage – needle holes* – Fabric is damages around the corners of the embroidery, normally caused by not using the correct type and size of needle, putting too many stitches in the same location, and not tearing tear away backing properly, allowing the fabric to be damaged as the stitches are pulled out. This can be corrected by digitizing properly, reducing the stitch count in the corners, using the correct type and size of needle and a ball point needle as small as possible.
- *Fabric grin through or gapping* – Fabric being seen through the embroidery design either in the middle of the pattern or on the edge can be corrected by digitizing properly (using appropriate underlay stitches, increasing stitch density, using different fill stitch pattern or direction, or compensating for "Pull" of thread by overlapping fill and satin border stitches), and using appropriate topping.
- *Missed trims* – The threads are left on the embroidery pattern between images or lettering. Thread trims are digitized when changing colours and when moving from one location to other using "jump" stitches. This can be corrected by digitizing properly (using appropriate number of trims, appropriate tie-off stitches, or replacing

trimming knives when necessary) and hand trimming missed trims using trimming snips.

- *Poor coverage* – Poor coverage or poor stitch density is where the stitch density is not thick enough and one can see through the embroidery stitching. This can be corrected by digitizing properly (using appropriate stitch selection, more stitches, and underlay stitches) and using appropriate backing and topping.
- *Poor hooping* – The fabric around the embroidery looks distorted and does not lay flat. It can be corrected by using appropriate backing and topping, making sure sewing operators hoop the garment properly without stretching the fabric too much prior to putting it in the hoop, and pressing or steaming hoop marks.
- *Poor registration* – Poor registration is where the stitches and design elements do not line up correctly. The embroidery sewing process sews diffgrent colours at different times. If the fabric shifts while one colour is being sewn, then poor registration will occur when the next colour is sewn. Sometimes it is difficult to tell the difference between poor registration, poor digitizing, and fabric "grin-through" or "gapping" due to thread "pull". Generally can be corrected by: Digitizing properly (using appropriate underlay stitches) and Hooping properly (using correct backing to prevent excessive material flagging).
- *Poor stitch balance* – This is where white bobbin thread shows on the topside of the embroidery. Ideally, the needle thread should be held on the underside of the seam, and not ever be pulled up to the topside. Proper stitch balance can be checked on the underneath or backing side of the embroidery by looking for 2/3 needle thread to 1/3 bobbin thread on Satin stitches. This can be corrected by using quality embroidery needle thread, quality pre-wound bobbins and setting machine thread tensions correctly.

15.24 Seam quality defects

- *Improper stitch balance – 301 lock-stitches* – The loops are seen either on the bottom side or topside of the seam. This is prominent with different coloured needle and bobbin threads and also, this defect comes where the stitch is too loose. To overcome this problem use a quality thread with consistent frictional characteristics, properly balance the stitch so that the needle and bobbin threads meet in the middle of the seam. Always start by checking the bobbin thread tension to make sure it is set correctly, so that the minimum thread tension is required to get a balanced stitch.

- *Improper stitch balance – 401 chain stitch* – Where the loops on the bottom-side of the seam are inconsistent and do not appear uniform. To overcome this use a quality thread with consistent frictional characteristics, properly balance the stitch so that when the looper thread is unravelled, the needle loop lays over half way to the next needle loop on the underside of the seam.
- *Improper stitch balance – 504 over-edge stitch* – Where the needle loop is not pulled up to the underside of the seam and the "purl" is not on the edge of the seam we get over edge stitch. To overcome this use a quality thread with consistent frictional characteristics and properly balance the stitch so that when the looper thread is unravelled, the needle loop lays over half way to the next needle loop on the underside of the seam.
- *Needle cutting on knits* – The needle holes appear along the stitch line that will eventually turn into a "run". This defect is caused by the needle damaging the fabric as it is penetrating the seam. Make sure the proper thread size and needle type and size are being used for the fabric, the fabric has been properly stored to prevent drying out and finished properly and check with your fabric manufacturer.
- *Open seam – seam failure – fabric* – Open seam is where the stitch line is still intact but the yarns in the fabric have ruptured. Solutions are reinforcing stress points with bartacks. Make sure the bar tacks are the proper length and width for the application, make sure the patterns has been designed for proper fit, make sure the ideal seam construction is being used, and contact your fabric supplier.
- *Open seam – seam failure – stitch* – Where the threads in the seam have ruptured leaving a hole in the stitch line, caused by improper stitch for application, inadequate thread strength for seam and not enough stitches per inch. The solutions are using a better quality sewing thread, the proper size thread for the application. For knit fabrics, check for "stitch cracking" caused by not enough stitches per inch, improper seam width or needle spacing for application, improper stitch balance and improper thread selection.
- *Puckered seams – knits and stretch woven* – Puckered seam is where the seam does not lay flat after stitching mainly due to too much stretching of the fabric while sewing. The solutions include setting the sewing machines properly for the fabric if sewing machines are equipped with differential feed, using minimum presser foot pressure during sewing and adopting correct handling techniques.
- *Excessive seam puckering – woven* – The seam does not lay flat and smooth along the stitch line. The reasons may be 'feed puckering', where the plies of fabric in the seam are not being aligned properly during sewing, 'tension puckering' where the thread has been

stretched and sewn into the seam causing the seam to draw back and pucker and yarn displacement or 'structural jamming' caused by sewing seams with too large of thread causing displacement of yarn in the seam. To avoid this use the correct thread type and size for the fabric, (In many cases, a smaller, higher tenacity thread is required to minimize seam puckering but maintain seam strength), sew with minimum sewing tension to get a balanced stitch, make sure that machines are set up properly for the fabric being sewn and check for proper operator handling techniques.

- *Ragged/Inconsistent edge* – Over-edge or safety stitch seams are where the edge of the seam is either extremely "ragged" or "rolls" inside the stitch. To avoid this sharpen the sewing machine knives and change regularly, adjust the knives properly in relationship to the "stitch tongue" on the needle plate to obtain the proper seam width or width bite.

- *Re-stitched seams / broken stitches* – This is the defect where a "splice" occurs on the stitch line. This is highly objectionable in top stitching. It is caused by thread breaks or thread run-out during sewing, or cut or broken stitches during a subsequent treatment of the finished product (i.e., stone washing). To avoid this use a better quality sewing thread. This may include going to a higher performance thread designed to minimize sewing interruptions. Ensure proper machine maintenance and sewing machine adjustments. Make sure sewing machines are properly maintained and adjusted for the fabric and sewing operation. Observe sewing operators for correct material handling techniques.

- *Re-stitched seams in jeans* – If there is a splice on stitch line and occurs on top stitching, it is objectionable. It may be caused by breaks or thread run out during sewing, or cut or broken stitches during a subsequent treatment of the finished product. The solutions include using better quality sewing thread, ensuring proper machine maintenance and adjustments of sewing machine and observing sewing operators for correct material handling techniques.

- *Broken stitches – needle cutting in jeans* – When a thread is being broken one seam crosses over another seam resulting in stitch failure like bartacks on top of waistband stitching, seat seam on top of riser seam. Using the proper thread and maintaining the proper stitch balance can minimize broken stitches due to needle cutting. Use of higher performance perma core or D-core thread, using a larger diameter thread on operations where the thread is being cut, making sure the proper stitch balance is being used, using needles with the correct point and changing the needles at regular intervals on operations are the remedies.

- *Broken stitches – abrasion in jeans* – The thread on the stitch line is broken during stone washing, sand blasting, hand sanding, etc. Broken stitches must be repaired by re stitching over the top of the stitch-line. The prevention can be done by use of higher performance perma core or D-core thread, use of larger diameter thread on operations where excessive abrasion is occurring (e.g. waist band), ensuring that stitches balance properly, using air entangled thread in the looper due to its lower seam profile making it less susceptible to abrasion (in yoke, seat and waistband seam) and monitoring the finishing cycle.
- *Excessive seam grin* – Excessive grin is where the stitch balance is not properly adjusted (stitch too loose) and the seam opens up. To check for seam grin, apply normal seam stress across the seam and then remove the stress. If the seam remains opened, then the seam has too much "grin through". To correct, readjust the sewing machine thread tensions so that the proper stitch balance is achieved. Too much tension will cause other problems including seam failures (stitch cracking), excessive thread breakage, and skipped stitches.
- *Seam failure* – Seam Slippage is where the yarns in the fabric pull out of the seam from the edge. This often occurs on fabrics constructed of continuous filament yarns that are very smooth and have a slick surface and in loosely constructed fabrics. To avoid consider changing the seam construction to a French seam construction, increase the seam width or width of bite, optimize the stitches per inch and contact your fabric supplier.
- *Skipped stitches* – This is where the stitch length is inconsistent, possibly appearing as double the normal stitch length; or that the threads in the stitch are not properly connected together. It is caused by the stitch forming device in the sewing machine missing the thread loop during stitch formation causing a defective stitch. On looper type stitches, this will allow the stitch to unravel causing seam failure. To avoid this use a better quality sewing thread, ensure proper machine maintenance and sewing machine adjustments, make sure that sewing machines are properly maintained and adjusted for the fabric and sewing operation. Observe sewing operators for correct material handling techniques.
- *Skipped stitches in jeans* – Where the stitch forming device misses the needle loop or the needle misses the looper loop. Skips are usually found where one seam crosses another seam and most of the time occurs right before or right after heavy thickness. The solutions are using core-spun thread, minimum thread tension to get a balanced stitch, the ideal foot, feed and plate that help to minimize flagging, training sewing operators not to stop on the thickness, making sure

the machine is feeding properly without stalling and the machine is not back-feeding.

- *Unravelling buttons* – This is where a tail of thread is visible on the topside of the button and when pulled, the button falls off. To avoid this use a quality sewing thread to minimize skipped stitches, specify attaching the buttons with a lockstitch instead of a single thread chain stitch button sewing machine.

- *Broken stitches – due to chemical degradation in jeans* – The thread in seam is degraded by the chemicals used during laundering resulting in loss or change of colour and seam failure. The solutions include using higher performance Perma Core NWT that has higher resistance to chemical degradation. It is recommended to go for larger thread sizes when the denim garments are subjected to harsh chemical washes. Ensure proper water temperature and pH levels, and proper amount and sequence of chemical dispersion as per guidelines and proper rinsing and neutralizing. Monitor the drying process, cycle time, and temperature

- *Unravelling seams in jeans* – Generally occurs on 401 chain stitch seams where either the stitch has been broken or a skipped stitch has occurred. This will cause seam failure unless the seam is re stitched. The solutions include using a high performance Perma Core or D-Core thread that will minimize broken stitches and skipped stitches, ensuring proper maintenance and adjustments of sewing machine and training sewing operators for correct material handling techniques.

- *Sagging or rolling pockets* – Where the pocket does not lay flat and rolls over after laundering. The solutions include making sure the sewing operators are not holding back excessively when setting the front pocket, the hem is formed properly and that excessive fabric is not being put into the folder that will cause the hem to roll over. Ensure that pocket is cut properly and pocket curve is not too deep. Use a reinforcement tape on the inside of the pocket that may help prevent the front panel from stretching along the bias where the front pocket is set. Select fabric construction as the type and weight of fabric also can contribute for this.

- *Ragged / Inconsistent edge* – This is where the edge of the seam is either extremely "Ragged" or "Rolls" inside the stitch. To avoid this make sure the sewing machine knives are sharpened and changed often. The knives should be adjusted properly in relation to the "Stitch Tongue" on the needle plate to obtain the proper seam width or width bite.

- *Wavy seam on stretch denim* – Where the seam does not lie flat and is wavy due to the fabric stretching as it was sewn or during subsequent laundering and handling operations. To avoid this use minimum presser foot pressure. Instruct sewing operators to use proper handling techniques and not stretch the fabric as they are making seam. Where available, use differential feed to compensate for the stretch of the fabric.

- *Ropy hem* – Ropy hem is where hem is not laying flat and is skewed in appearance, usually caused by poor operator handling. Sewing operator should make sure they get the hem started correctly in the folder before they start sewing and should not hold back excessively as the seam is being sewn. Use minimum roller or presser foot pressure.

- *Twisted legs in jeans* – Twisted leg is where the side seam twists around to the front of the pant and distorts the appearance of the jeans, usually caused by poor operator handling. To avoid this sewing operator should match the front and back properly so they come out the same length. Notches might be used to ensure proper alignment. Ensure that operator does not trim off the front or back with scissors to make them come out the same length. Make sure the cut parts are of equal length coming to the assembly operation. Check fabric quality and cutting for proper skew. Make sure the sewing machine is adjusted properly for uniform feeding of the top and bottom plies.

- *Disappearing stitches in stretch denim* – Where the thread looks much smaller on seams sewn in the warp direction than in the weft direction of the fabric. To prevent this use a heavier thread size on top stitching [120 to 150 Tex], go to a longer stitch length [from 8 to 6 S.P.I] and make sure the thread tensions are as loose as possible so the thread sits on top of the fabric rather than burying in the fabric on seams sewn in the warp.

- *Thread discolouration after laundry in jeans* – The thread picks up the indigo dyes from the fabric giving the thread a 'dirty' appearance. A common discoloration would be the pick up of a greenish or turquoise tint. The main reasons are improper pH level, improper water temperature, improper chemical selection and shortcuts on wash methods. The solutions for this are using thread with proper colour fastness characteristics, correct pH level and low water temperature during laundry, using the proper chemicals and laundry cycles, and using denimcol PCC or similar additive in wash. Do not over load washers with too many garments at one time.

- *Poor colour fastness after laundry* – The thread does not wash down consistently in the garment or changes to a different colour all together. The normal reasons are mixing threads in a garment, using threads with different colour fastness and not doing preproduction testing. To avoid this use thread with proper colour fastness characteristics, use threads from same thread supplier and do not mix threads in a garment. Always do preproduction testing on denim garments using new colours to assure that they will meet the requirements. Make sure sewing operators select thread by type and colour number and do not just pick a thread off the shelf because it looks close in colour.

15.25 Fitting related defects in garments

- *Improper fit* – Fitting is one of the important criteria for consumers in their buying decision. Variation in dimensions and improper fitting are the normal complaints in readymade garments. The solutions include collecting data on age, body structure and ethnicity. Data can be colleted by sample survey method that can represent the population as a whole. Appropriate statistical tools can help in getting range and variation of sizes found in people. Adopting technologies such as automated measurement and 3D body scanning facilitate more effective and affordable data collection for garment manufacturing companies. The effectiveness of sizing system is highly dependent on skills of patternmakers and graders in identifying, defining and manufacturing the type of fit appropriate for the target market. Some of the tools and strategies to facilitate this are target body scanner, market surveys, wear testing, and virtual fit assessment. Materials with stretching characteristics can fit a wider range of body.
- *Improper size labels* – Many times it is seen that consumers get confused with the sizes. They are not able correlate the numbers with body measurement, so they prefer taking trials of various sizes of same item. The reason behind this is non standardization of label sizes among manufacturers. To get rid of this problem, manufacturers are using terms such as slim, classic, relaxed fit. Use of Internationals Standards for dress size and educating the users on the Standards adopted could solve the problem to a certain extent.
- *Cutting inaccuracies* – Factors that cause cutting inaccuracies are wide or vague lines on the marker, imprecise following of lines on the marker, variation in the cutting patch, shifting of the spread or block, allowing fabric to bunch up or push ahead of the knife, using improper equipment and using improper cutting sequence as parts are cut.

References

1. Don, B. *Great Leadership Depends*, Don@LeadWell.com., www.LeadWell.com.
2. The Characteristics of a Leader: Demonstrating Good Leadership Skills http://www.coach4growth.com/good-leadership-skills/characteristicsofaleader.html
3. Saad Deti Yasha Shikhare – By Prof S. D. Mahajan (Marathi)
4. Guidelines for Process Management in Textiles – By B. Purushothama – CVG Books Publications
5. Winning Strategies – By B. Purushothama – Pubadchi Publications
6. Manku Thimmana Kagga – Br Dr. D. V. Gundappa (Kannada)
7. Huchchuraamana Muktakagalu – By B. Purushothama (Kannada)
8. Motivating a Team – Dale Carnegie Training – Dale Carnegie & Associates
9. A Practical Guide for Quality Management in Spinning – By B. Purushothama, Woodhead Publishing India
10. Quality Systems for Garment Manufacture – By Ellis Developments Ltd. Nottingham, UK
11. Q.C. Tool Guide book by C.I.I – 1995
12. Embroidery Defects Checklists – American and Efird Inc.
13. Defects in wet processing, Causes and Remedies by Siddeshwar Gaddam and Digambar Mirajkar, Solapur Textile Directory 1999-2000
14. Visual Management through Five S: A Japanese tool of Kaizen – By Shyam Talawadekar
15. Basics of Quality Leadership – By Tata Quality Management Services
16. Course curricula for Short term courses based on Modular Employable Skills, Sector – Textiles, Director General of Employment and Training, Ministry of Labour and Employment, Government of India
17. Course curricula for Short term courses based on Modular Employable Skills, Sector – Garment, Director General of Employment and Training, Ministry of Labour and Employment, Government of India

18. Quality People – Key to Excellence – 6th Asian Network for Quality – Bangkok 2008.
19. Measuring Customer Satisfaction – Approaches for Getting Reliable Information for Textile and Garment Industries – By B. Purushothama – Fibre2fashion.com
20. Fairchild's Dictionary of Textiles, 7th Edition, Fairchild Publications, New York, 2000
21. The Change Agents Handbook – David W. Hutton
22. Quality Management in Garment Industry by B. Purushothama, ISTE Publication 2007
23. Tablets Published by Textile Association India on different processes of Textile Mills
24. Overcoming Resistance to Change – By Dr. Joel. R. DeLuca
25. Change Management by Allan Chapman – Business Balls.com

Index